AGRICULTURAL PROPERTY RIGHTS

AGRICULTURAL PROPERTY RIGHTS

Nekesah T. Wafullah

Delve
PUBLISHING

www.delvepublishing.com

Agricultural Property Rights

Nekesah T. Wafullah

Delve Publishing
4164 Lakeshore Road
Burlington, ON L7L 1A4
Canada
www.delvepublishing.com

Email: orders@arclereducation.com

ISBN: 978-1-77956-483-2 (Hardcover)

ABOUT THE AUTHOR

Nekesah T. Wafullah is a skilled agriculture expert with extensive knowledge in agricultural energy value addition products, agricultural business management services, project management, various forms of fertilizer, their production, sales, marketing aspects and application regimes; cross border fertilizer trade policies; youth and women empowerment and volunteerism. She is adept at project planning and management as well as creating simple solutions to complex problems. She has experience within agricultural markets in Kenya, Zambia, Malawi, Tanzania, Rwanda, the Democratic Republic of Congo (DRC)- Bukavu and Lubumbashi and Uganda. She mentors high school and college students and advocates for better performance in Agricultural science. During her free time, she loves editing books, watching movies, cooking, baking, networking, reading and dancing. Nekesah holds an M Sc. In Agricultural and Applied Economics degree from the University of Nairobi with a major in International Trade and Policy.

TABLE OF CONTENTS

LIST OF FIGURES

LIST OF TABLE

LIST OF ABBREVIATIONS

AFMA	Americans for Medical Advancement
ALF	Animal Liberation Front
ASSINSEL	International Association of Plant Breeders for the Protection of Plant Varieties
AVC	Antivivisection Coalition
B.C.E.	Before the Common Era
BT	Bacillus Thuringiensis
CBD	Convention on biological diversity
CBOs	Community–Based Organizations
CEED	Center for Entrepreneurship and Economic Development
CFI	Cruelty Free International
CGIAR	Consultative Group on International Agricultural Research
CIPE	Center for International Private Enterprise
CPR	Common Property Regimes
DNA	Deoxyribonucleic Acid
EBA	Ecosystem-Based Adaptation
EC	European Commission
ECJ	European Court of Justice
EPO	European Patent Office
EU	European Union
FAO	Food and Agriculture Organization
FDI	Foreign Direct Investment
FLOE	For Life on Earth
FTA	Free trade Agreement
GMO	Genetically Modified Organisms
GURTs	Genetic use restriction technologies
HSUS	Humane Society of the United States
HYVs	High Yielding Varieties
IGC	International Growth Centre

IKSs	Indigenous Knowledge Systems
IP	Intellectual Property
IPRs	Intellectual Property Rights
IRRI	International Rice Research Institute
MC	Marginal Cost
MTAs	Material Transfer Agreements
NASAA	North American Securities Administrators Association
NCBI	National Center for Biotechnology Information
NGOs	Non-governmental organizations
NISCO	National Informal Sector Coalition
OAU	Organization of African Unity
OECD	Organization for Economic Co-operation and Development
OSSI	Open-Source Seed Initiative
PBR	Plant Breeders' Rights
PBRA	Plant Breeders' Rights Act
PCT	Patent Cooperation Treaty
PDO	Protected Designations of Origin
PETA	People for the Ethical Treatment of Animals
PGI	Protected Geographic Indications
PGRFA	Plant Genetic Resources For Food And Agriculture
PVR	Plant Variety Rights
R&D	Research and Development
RNA	Ribonucleic Acid
RuR	Round Up Ready
TAD	Technical Assistance Database
TRIPS	Trade Related Aspects of Intellectual Property Rights
UK	United Kingdom
UNC	University of North Carolina
UNCHS	United Nations Centre for Human Settlements
UNEP	United Nations Environment Program
UNESCO	United Nations Educational, Scientific and Cultural Organization
UPOV	Union for the Protection of New Varieties of Plants
USDA	United States Department of Agriculture

VAP	Value-Of-Average-Product
VMP	Value of Marginal Product
WHO	World Health Organization
WIPO	World Intellectual Property Organization
WTO	World Trade Organization

ACKNOWLEDGMENT

The book is the result of significant effort and time. Its completion would not have been possible without the participation and assistance of many individuals whose names may not all be mentioned. Their contributions are sincerely appreciated and gratefully acknowledged. However, I would like to extend my deepest appreciation and gratitude, particularly to the following: Arcler Education, Inc. and James M. Kihungi for their endless support and their kind and understanding spirit throughout this endeavor. To all relatives, friends, and others who, in one way or another, offered their support—whether morally, financially, or physically—I extend my heartfelt thanks.

PREFACE

Agriculture requires stable access to resources like land and water, as well as the right to claim ownership of livestock. Thanks to property rights, this is feasible. Property rights must be precisely defined and zealously upheld since they are the foundation of our society. When it comes to the owned object, property rights usually encompass a broad range of unique freedoms, rights, privileges, and immunities. The concept of property rights and whether they should exist at all are highly contentious topics. More specifically, the question is what broad guidelines should be followed when transferring property rights. These usage guidelines give birth to various property regimes, such as open access, private ownership, and government control over the resource base. Every regime has pros and cons of its own. The regime with the maximum efficiency has been shown to be the private one. Some interested parties may believe that the private system is better because of its efficiency and economic returns, but this is untrue, and this strategy should not be used to support a regime change. However, to guarantee economic growth and property security, a more extensive strategy should be used. Additionally, some things, including the atmosphere and oceans, cannot be included in the regime of private property.

To promote positive behavior among entrepreneurial individuals, a strong property rights framework works more than any other single component. People who own property are more willing to put in effort and accept risks than they otherwise would. Possession of real estate means opportunity, accountability, and financial independence for individuals. These characteristics benefit society by fostering innovation, investment, the potential for extensive trade, and even better government. Property creates the basis for development, liberty, and general well-being by influencing people's incentives and supporting societal institutions. Therefore, nations that lack strong property rights and other essential market economy institutions experience slow or negative economic growth, while those that do enjoy prosperity. The agriculture sector has a centuries-old connection to property rights. It would be difficult for farmers and ranchers to regulate how agricultural resources are used if they did not have acknowledged property rights over them, either through title or a lease. Control is essential for granting ownership of livestock and safe, steady access to scarce, valued resources like land. Farmers and ranchers are more prepared to take chances and experiment with new methods if secure and stable access or ownership has been established. Property security in agriculture is mostly through titling and registration. In addition to tangible property, intellectual property also are very important in promoting agricultural growth. Intellectual property (IP) safeguards any legal rights that belong to the original creator of the notion and guards against theft and misuse. The rights that safeguard an intellectual concept or the goods and services that are derived from it are known as intellectual property rights, or IPRs. Intellectual property takes four forms Trade secrets, trademarks, copyright, and patents which are monitored by organizations such as WIPO and STRIP. Despite critique to IPR in agriculture, they greatly contribute to innovation towards food security and biodiversity.

–Author

Chapter 1

Property Rights

Contents

OVERVIEW

Since property rights form the cornerstone of our society, it is critical that they be clearly defined and vigorously protected. In addition to elevating the individual owner, having control over one's property enables society to flourish through the use of that property by the owner. The agriculture sector is one that has relied on property rights since its beginning. The ability to claim ownership of cattle and steady access to resources like land and water are prerequisites for agriculture. Farmers and ranchers benefit from property rights by having the required access, confidence, stability, lower risk, and the chance to earn a living using their own resources. Farmers and ranchers can produce food and support the economies of governments and nations when they are protected by robust property rights. Several states in the United States passed farm protection laws, also known as "ag-gag laws," to safeguard farmers' property rights so they could continue to produce and contribute. These rules criminalize a number of actions, such as lying to secure employment or access to a private agricultural facility in order to film a farmer's operations and publish the footage publicly to defame the operator. Farmers may experience abuse, death threats, and financial instability if recordings like this are made public. In order to prevent farmers from being exploited on their own private property, farm protection laws were passed; however, courts have lately declared several of these regulations unconstitutional on the grounds of First Amendment freedom of speech. By undermining the security of a farmer's property right to keep others off his land, the invalidation of these regulations could have a serious effect on the agriculture sector. Ag-gag laws are intended to prevent undesirable individuals from using false identities to obtain access to private agricultural facilities and to cease undervaluing and marginalizing the work of ranchers and farmers. Landowners are more likely to face ongoing threats from unauthorized individuals in the absence of these rules. Unsecured property rights cause farmers and ranchers to lose confidence, take fewer chances, adhere to outdated practices and technologies, and not pursue new and superior methods if they may be easily thwarted.6. State and national economies will be affected by a stagnant agricultural sector, society will suffer from a decrease in the quantity and quality of food products, and the environment will be harmed by the lack of adoption of more sustainable practices due to fear of the consequences of activist interference in the agricultural sector. The establishment and maintenance of a robust agriculture sector—which can only be achieved through safe and stable property rights for farmers—is essential to halting this impact's cascade. Property rights should be fiercely maintained and upheld wherever possible since they have a long history of being extremely important in our society and around the world.

1.1. PROPERTY RIGHTS CONCEPT

It may appear that defining property rights requires defining two distinct concepts—the notion of property and the concept of a right—but it should be highlighted that the first explanation is necessary for the second. This is due to the fact that property rights are a prerequisite for the normative concept of property. This point may be overlooked because the term "property" and other ownership signifiers can be used to identify entities that belong to someone (e.g., "This is my book, not yours") as well as to establish, claim, challenge, or report ownership relations (e.g., "Could you give me my bag, please?"). In the latter case, the normative nature of the ownership relation is still present. Nonetheless, all usages of the term "property" assume its normative aspect. Only when someone has a legitimate claim to property can an entity be considered property. Therefore, the broad idea of a right should be the starting point for a definition of property.

1.1.1. Concept of Rights

Normative relations are established by rights. These relationships consist of at least:

- Someone who is the rights holder;
- Something to which the rights holder has the right; and
- Someone against or with respect to whom the rights owner has the right.

Relationships based on rights are always maintained by some foundation for justification. At best, it would be bluffin□to assert a ri□ht to X without havin□ the ability to respond to any potential challenges to that right. Moral arguments (for moral rights) or legislative decisions (for positive legal rights) might serve as the foundation for the rationale. Their social efficacy may be □uaranteed by punishments in addition to the force of reason, as is frequently the case with legal rights. Some philosophers even go so far as to say that the employment of coercive methods to compel obedience is a necessary component of rights since it makes them valid. The normative power of a rights claim in support of X is unquestionably □reater than the ar□ument that X would be beneficial, or even beneficial for all of us, even thou□h this criterion may be overly narrow. In a metaphor that has been frequently used ever since, Ronald Dworkin stated that individual rights take precedence over arguments in support of group objectives. Donatin□blood could be beneficial to society, for instance. But no one has the right to make you compromise your bodily integrity. Instead, your right will take precedence over opposing evaluative factors, provided that these factors do not also support a conflictin□ri□ht. Utilitarian explanations are not excluded by this concept of rights. It undoubtedly permits rule-consequentialist defenses of a rights system. However, it appears that pure act-utilitarianism only recognizes one right, which is the right of any moral subject to have her personal utility taken into account while determining her overall utility. No other individual claim could ever acquire the normative force of an individual right in Dworkin's view as it would always be superseded by a potential increase in □eneral benefit. Wesley N. Hohfeld is credited with developin□a □enerally reco□nized classification of ri□hts. He contends that we ou□ht to differentiate rights—what Hohfeld refers to as claim rights—from liberties, powers, and immunities. Other parties are always obli□ated to fulfil a claim ri□ht. Both positive and negative claim rights are possible. If a right holder has a claim right to Y against another person, the other person either has a positive claim right if she helps the right holder obtain Y, use Y, do Y, etc., or she has a negative claim right if she simply has an obligation to refrain from interfering with the right holder's possession of Y, use of Y, doing Y, etc. Regarding freedoms, powers, and immunities, a liberty is the lack of conflictin□claim ri□hts; yet, as lon□as it is not accompanied by a claim right, it does not impose any obligations, either positive or negative, on other parties. Immunities are claim rights that shield one's rights from the authority of others, while powers are second-order rights that can chan□e first-order ri□hts. Let's examine the ideas of human ri□hts and natural rights to wrap up our broad thoughts on the concept of rights: Either the □roup of people who have ri□hts or their content can be used to define human rights. The first perspective is that human ri□hts are those that one possesses simply by virtue of bein□human. The second, more forceful definition of human rights as those that ensure the fundamental conditions for an existence that can be genuinely (in an evaluative sense) referred to as human may be compatible with this definition, but it is not necessarily implied. In order to make any moral right seem natural, natural rights are occasionally confused with nonpositive rights. However, it would seem more appropriate to limit the class of natural rights to those rights (if any) that may be primarily supported by naturalist arguments or, even more narrowly, by some allusion to the essence of humanity.

1.1.2. Philosophy of Property

The allocation of advantages, rights, and obligations related to goods in a society is the focus of property philosophy. The majority of prominent Western political philosophers, such as Plato, Aristotle, Hume, Locke, Mill, and Marx, have developed theories of property rights and ownership, including the boundaries and rationale of property. Property rights philosophy is intimately related to distributive justice, the function of the state, and the basic conflicts between modern liberal and socialist political thought. In addition to critically analyzing the arguments made in favor of property systems, philosophical interest in property and property rights entails conceptual study of what property, ownership, and property rights are. This entry discusses the definition of property, property rights, and ownership before looking at the historical and intellectual contexts in which theories of property are developed, summarizing opposing viewpoints on property justification, and finally discussing current and future property-related arguments in political philosophy and philosophy of law.

1.1.2.1. The Term 'Property'

Relationships are arguably the best way to understand property. Property rights include relationships between individuals regarding advantages and obligations that are frequently (though not always) connected to specific items in the environment. However, in everyday discourse, "property" is frequently used to refer to the things that are the subject of property relations. Philosophical discussions of property focus on property understood as social, legal, and moral relations with regard to objects, benefits, responsibilities, and their control, within specific legal and social contexts, even though it is common to talk about property in relation to objects in the world. The legally acknowledged allocation of rights, obligations, and advantages related to property (both tangible and intangible) is the subject of property law. However, legally acknowledged conceptions of the boundaries and scope of property rights do not bind social and moral connections of property. Each jurisdiction has a different set of legally recognized property rights and obligations. Likewise, moral and social property rights differ throughout history and throughout cultures. Determining whether a specific claim to a benefit or attribution of obligation is better described as a property right or as another type of legally or socially accepted benefit or responsibility can be challenging. It may be impossible to distinguish precise differences between property, contract, and tort bases for legal entitlements or legally protected interests in jurisprudence. Philosophical debates on property are thus inextricably linked to more general debates about distributive fairness, social justice, and the role of the state in defending or advancing particular types of interests. The scope and boundaries of property rights differ in three ways: the scope of entities that can possess specific property rights (such as individuals, corporations, communities, and the state); the scope of specific property rights over an object that are recognized (such as whether property can be bequeathed upon death or whether property rights can be leased to others); and the scope of objects over which property rights can be held (such as whether water, air, or titles can be held as property). Even though the state may have good moral reasons to restrict property rights in the public interest (for instance, banning the sale of human beings or banning the purchase of insurance against certain legal responsibilities like legal culpability for criminal conduct), there is a presumption that anything could be considered property and that anything that is property could be alienated in capitalist states. The types of property that can be privately owned may be restricted in pre-capitalist and socialist states (for instance, land, factories, or other means of production may be owned by the state or collectively), and there may be a presumption that the state has an interest in restricting private property rights in order to redistribute benefits and burdens through measures like taxation, price control, restrictions on the right of sale, and wage control.

1.1.2.2. Property as a 'Bundle of Rights'

The term "property" is commonly used in conceptual analysis to refer to a collection of rights, or rights and obligations. The entire range of rights and obligations enjoyed by a private owner of an item is where the specification of the bundle begins in liberal democratic philosophy and jurisprudence. These are interpreted to entail, at the very least, the Hohfeldian rights of exclusion, use, alienation of property, and exemption from expropriation (Hohfeld, 1919). Although theorists have offered several methods for listing the various facets (rights and obligations) of ownership, the majority are currently greatly impacted by Honore's (1961) list of the 11 "incidents" that collectively make up "full, liberal ownership." These include the rights to use, hold, manage, earn, invest, secure, and have no term or residuality, as well as the obligations to refrain from harmful use and to be held accountable for execution. Though it is meant to apply to intangible property like future production, labour control, and copyright, Honore's theory is most easily applied to property in material items. Therefore, these types of property are not intended to be excluded by the terms "thing" and "object over which property rights are held." The fundamental rights that collectively constitute ownership may be characterized in a different way according to Honore's research; for instance, more precise distinctions may be drawn between the right to capital through the sale of property and the right to consume it. He also acknowledges that the breadth of property rights that an individual needs possess to be considered the owner of a certain object (like a car) may differ from the requirements for owning another type of object (like a pharmaceutical firm). Other theorists, like as Becker (1977), offer unique, less thorough explanations of property that are based on Honore's list.

1.1.2.3. Ownership

Philosophical discussions of property are frequently interpreted as discussions of "ownership" within the context of liberal capitalism. When it comes to other property rights, ownership may be viewed as a sort of legal trump, but this isn't always the case for other moral or social claims. The rights an owner has over an object, which nonowners are not allowed to meddle with, have been the subject of numerous authors who have attempted to examine the ideas of property and ownership as social and legal institutions. When someone has "the greatest possible interest" in something, they are said to be in a relationship of "ownership," which is typical in a private property system (Honore, 1961). In other words, the title is awarded to the individual who possesses the greatest variety of property rights over the item. Perhaps it would be more accurate to state that the owner is the individual who possesses the most important property rights, either legally or socially, over a given type of object. Depending on the type of property owned and the potential uses of those rights, different rights will be deemed more important. For example, the owner of an industrial business may be identified by their control over production and the income from profits, whereas the owner of a retail space may be identified by their control over access and use; the owner of personal property (like a car) may be identified by their possession and use rights. Legally speaking, the person who is considered the "owner" of the thing has a stronger claim than any other property rights holder. However, there are other situations in which attempting to ascertain who is entitled to ownership is ineffective because property rights can be divided and restricted in a plethora of ways. Furthermore, while talking about private property, the phrases "owner" and "ownership" most naturally come up. Rather than trying to define ownership relations, it is often more beneficial to refer to the range of property relations acknowledged within the theory when defending property theories where the relative merits of private, communal, state, or mixed systems of property are to be evaluated.

1.1.2.4. Property Systems

The legal and social structures pertaining to property in a culture are commonly classified into three types: common, collective, and private property systems. Private, collective, and common property systems are distinguished by the distinct "organizing idea" that each type of system is based on, according to Waldron (1988), who defines property systems as "rules governing access to and control of material resources." The rules that regulate access to and control of resources in a private property system are structured on the notion that resources are, in general, distinct things that are assigned to and, so, belong to a certain individual. The organizing principle or "social rule that the use of material resources in particular cases is to be determined by reference to the collective interests of society as a whole" is what defines collective property systems. The premise that "each resource is in principle available for the use of every member alike" serves as the organizing principle for the rules governing the distribution of resource rights in common property systems (Waldron, 1988). Waldron distinguishes between different types of property systems based on their organizing ideas; nonetheless, his notion of an organizing idea for property systems is not intended to be reductive. The majority of real-world property systems include components of each system: some resources, like sunlight, are considered common property, while others, like state-run utilities, are considered communal property. Other resources, like clothing, are owned privately.

1.1.3. Property Rights

Property rights are a complicated topic, if there is one uncontroversial statement regarding their interpretation. The majority of writers agree that the concept of property rights typically entails a wide range of distinct rights, freedoms, privileges, and immunities with regard to the owned item. Using Hohfeld's distinctions as a basis, Tony Honore′ highlights at least 11 elements or "instances" that combine to form the liberal concept of full ownership in his frequently cited examination of property rights:

- The (claim) right to possess – that is, the right to exert physical control over the object and to exclude others from it;
- The (liberty) right to use – that is, to make direct use of it for one's own personal purposes;
- The (power) right to manage it – that is, to decide how and by whom else it shall be used;
- The claim right to the income;
- The right to capital, which includes both the power to alienate and the liberty to consume or destroy the object;
- The (immunity) right to security – that is, the guarantee not to be expropriated (at least not without compensation);
- The (power) right of transmissibility; and
- The (immunity) right to the absence of term. Besides these different kinds of rights, the full notion of ownership is characterized by
- The prohibition of harmful use that restricts the ways an owner can use her property; (10) the liability to execution, which implies that an owner may have a duty to give his property away in order to repay a debt; and
- Rules for dealing with lapsed property rights.

It is clear, therefore, that not all 11 of these elements are always present in property rights, whether they are moral or legal. For instance, it is perfectly reasonable to claim that one can truly own a monument, even while property laws do not include the freedom to demolish or drastically alter the thing. Furthermore, a feature that appears to be crucial for a property right in one situation could be essentially incomprehensible in another. When it comes to tangible objects, for instance, the right to exclusive possession—that is, the right to physically control the object and to keep others from doing so—will typically be seen as a key component of ownership. However, the exact meaning of this right to exercise exclusive physical control may not even be evident in the context of intellectual property. We may observe that many of the elements in Honore's list allow for several specifications, which further complicates matters. For instance, you could have the right to erect a house on your property if you own it, but doing so will most likely result in additional expenses or advantages: A Neighbour might appreciate your home's protection from the wind or object to the shadow it creates. Your house may interfere with radio broadcasts or air traffic if it is hi□h enou□h. Therefore, the ri□ht to own land is typically governed by extensive laws regarding what you may do with it, such as whether you can build a house on top of it at all, how close it can be to the boundaries, how high it can be, etc. Although these requirements are unquestionably vital, it is crucial to remember that they might not be obvious from the idea of a property right alone.

There are a variety of responses to the Hohfeldian examination of the internal complexity of property rights, despite the fact that it has gained popularity in the field. First, proponents of the so-called disinte□ration perspective contend that a more precise terminology system should be used in place of "the" concept of property since it is not adequately explicit. A system of more comprehensive property rights (i.e., of property rights that include as many of its aspects as possible in the strongest possible interpretation) is at least initially preferable to a system of less comprehensive rights, according to those who defend the full liberal notion of ownership. They do this by either directly opposing the bundle of rights view and attempting to restore the traditional notion of a jus in rem or by defending the full liberal notion as a normative concept. Third, some ar□ue that it may still be reasonable to use a carefully defined notion of property rights or to divide this notion into a smaller number of more closely linked bundles (e.g., Christman proposes to distinguish between the two concepts of "control ownership" and "income ownership"). They do this by claiming that certain ownership-related rights bundles are more plausible or coherent than others. Even while these mostly address how to properly analyse the idea of property rights, they directly affect whether or not these ri□hts can be justified. Any defense or criticism of a property right would inevitably include defense or criticism of every facet of that right if it were conceptually necessary to interpret the concept of ownership in the sense of full liberal ownership, making any weaker sense of property rights inconsistent. Since the legitimacy of the more specific ri□hts or ri□hts bundles may rely on separate reasons, the disinte□ration perspective, on the other hand, or a view that permits various types of property-related bundles of rights, increases the need for explanation or criticism. There are some intriguing observations to be made solely on the level of conceptual analysis, and since the analysis of ownership does have some normative si□nificance, we may ask which of the aforementioned viewpoints is the most compelling. John Christman, for instance, draws our attention to an intriguing conflict that exists within the idea of full liberal ownership: The idea of an indissoluble unity of "full" ownership rights appears to be threatened by either granting owners the freedom to transfer the various aspects of their property rights separately or by si□nificantly restrictin□the types of transactions they are permitted to make with their property. However, the idea of "splitting" property rights into several rights bundles or even adopting the disintegration approach would necessitate si□nificant chan□es to many of our current applications of the notion, possibly resulting in further issues. This could hardly be considered an objection, then, if it were true that the conceptual underpinnings of our existing behaviours were seriously faulty. Ultimately, conceptual analysis alone cannot resolve the dispute over how to properly interpret the idea of property rights. An analysis of the term's de facto usage alone cannot explain any of the corresponding meanings of ownership. All of these views appear to be normative and revisionary to some de□ree. Therefore, while the possible justifications for the rights in question are directly impacted by the analysis of the concept of property rights, the opposite is also true: theories re□ardin□ the justification of property rights can support arguments against some interpretations of the concept of property rights while also making others tenable and practical. Thus, we can now at last discuss the rationale behind property rights.

1.2. APPROACHES AND JUSTIFICATIONS

The idea of property rights is hotly debated, as was previously indicated. The following lines will concentrate on the general question of the justification of property rights as such, despite the fact that there are numerous significant discussions regarding the moral significance of the legal institution of property rights in particular contexts (such as discussions regarding intellectual property, ownership of one's body or parts of it, or one's own genetic material). However, even this query permits specification: In its broadest sense, it is an issue of whether or not property rights should exist at all (or, conversely, whether or not some of the more specific ownership rights should exist). More precisely, it is a matter of what general rules ought to apply when assigning property rights. These two questions are somewhat related, depending on the theory in question. Common, collective, and private property are the three categories of property that can be distinguished, in accordance with Jeremy Waldron (notice that other terminologies are used in the literature). When it comes to common property, "objects" are accessible and usable by everyone. The purpose of claim rights, which safeguard common property, is to forbid private professions or other private uses that would deny others access to the advantages of the corresponding item. Ownership of an item that is utilised collectively, as determined by a communal decision, is known as collective property. Property rights are granted to people or legal entities, like corporations, in the context of private ownership. The fact that the institution of private ownership is at the centre of the debate over property rights is no accident. It is reasonable to presume that all decisions pertaining to collective and common property rights will need to be taken by the institutions of a legal political power in a collective manner. As a result, the idea of common and community property fosters rather than restricts the range of acceptable collective decision-making. However, when it comes to private ownership, property rights are viewed as personal liberties that limit the scope of political power. As a result, one of the mechanisms that keeps the private and political spheres apart and secure is private property rights. This is particularly true if these rights are viewed as pre-political moral rights—claim rights that are morally justified even before they are politically asserted—as opposed to just positive rights, which would only offer a limited defense against the political arena because property rights would once more be entirely dependent on political decision-making.

1.2.1. First Occupancy and Locke's Natural Rights Theory of Ownership

The first occupation hypothesis is the most basic form of a natural rights theory of private ownership. Its fundamental tenet is that a person has the right to claim ownership of something if she occupies it and it is not yet owned by another person. To avoid at least the most blatant paradoxical implications of this account, more details are needed. For instance, it is necessary to define what an act of occupation is in reality (rather than just declaratory) and what one can truly claim based on specific acts of occupancy (since everyone has a "foot in the universe," we could otherwise claim unlimited property rights). However, there are significant problems with the first occupancy argument even if it is modified by such specifications. Since first possession precludes the annexation of alien property, it does, of course, make some sense as a negative condition. Furthermore, if there are significant limitations and a system of property rights is already in existence, it can suggest itself as a particularly simple method of giving priority to ownership claims. However, the first occupancy method does not appear to be very promising as an attempt to establish property rights in the first place or to offer a stand-alone defense for particular claim rights to property. What normative relevance might occupations have? When someone steps on something, does that give them the right to make a genuine claim? There is a problem with any attempt to respond to this query: Either occupancy is viewed as a pre-social, entirely natural activity: Someone simply uses an object to perform something. Then, it's still unclear how a pre-social conduct like that could ever support a claim to rights against others. Or it is believed that the act of occupancy already carries a specific social connotation. However, the social significance of the act and its normative power would be established by a particular social rule rather than the act itself, such as the convention that activities of type X □ive rise to ownership rights of type Y. But this social norm would still need to be explained. Furthermore, it is a contingent reality that someone is the first to perform something. A person does not even have the opportunity to obtain anything herself if she is born after it was the object of first occupancy. Therefore, aside from prohibiting the annexation of foreign property, it appears unlikely that being first can add anything to support ownership. The most prominent natural rights theory of ownership was developed by John Locke, who is renowned for fusing a labour theory of property rights with the first occupation theory. Locke begins by assuming that although each man possesses property in his own person, God gave the world to all men as their common property. Among the latter is the right to one's own labour. A person has combined his labour with an item if he takes something that isn't already owned by someone else and changes it via his own efforts. The object becomes his property as it is now inseparable from something that belongs to him, as long as there is still enough for other people. Locke's reasoning is open to various interpretations, as has frequently been mentioned. It is possible to highlight at least three points: the part freedom plays in Locke's defense of private property rights, the concept of self-ownership, and the connection between productive labour and the desert.

1.2.2. Property Rights and Freedom

Locke's reasoning is interpreted as an application of his broad perspective on freedom and its boundaries in the first line. One must understand the act of working on and acquiring objects as a use of freedom. No one will be hurt if something is appropriated via labour and enough is left over for everyone. Consequently, it is acceptable to obtain property through employment. According to this perspective, the harm principle can be applied to the "Lockean proviso," which states that there is enough left in common. It goes without saying that the restriction significantly limits the potential uses of Locke's notion of legitimate acquisition. Because property rights are crucial, particularly when this clause is broken and resources are or may become limited. However, let's broaden our perspective and examine the overall relationship between freedom and property rights. Since property rights cannot be inferred solely from the right to liberty, it has been argued that they are different from other liberal rights. The premise is that a person's property rights directly increase her freedom at the price of other people's freedom. It has been countered that any (claim) right is equivalent to this. It is true that every individual right has associated responsibilities, both positive and negative, for those who do not possess the right. However, A's property right in X directly denies B the same rights, liberties, powers, and immunities in X that are granted to A by A's title of ownership, while B's right to life, freedom of conscience, or freedom of speech is not directly restricted by A's rights to life, freedom of religion, or freedom of speech; rather, it restricts B's freedom to kill or to suppress freedom of religion or speech. However, if it could be demonstrated that creating property rights is not a zero-sum game but rather eventually promotes freedom for all of us, then supporting property rights on the basis of a right to freedom would still be feasible. Despite their potential, such theoretical endeavour transform a simple natural rights argument into a consequentialist one, the viability of which depends on the soundness of intricate conceptual and empirical hypotheses. Hegel suggests another, perhaps more direct, link between property and freedom. He contends that the capacity of subjects to articulate and objectify their subjectivity in the material world is a prerequisite for positive freedom. Marx has adopted this viewpoint and, as a result, criticises worker expropriation as a type of alienation as well as economic injustice.

1.2.3. Self-Ownership

The concept of "self-extension" into the material world, which holds that things become your own by "mixing" them with your labour, is also present in Locke, in addition to Hegelian objective idealism. It has frequently been observed, meanwhile, that it is not at all convincing to infer property rights in things from those in the body and its operations. Even if the reasoning were definitive, it would not apply to land property, as early detractors pointed out. Furthermore, one would wonder why combining some of your property with some common property would make the part that is owned by everyone your private property instead of the other way around. The brute object must undoubtedly contribute something beneficial to your work, at least by serving as the raw material for your activity. One possible response would be that your labour, not the physical object, is what gives it value. Because something was taken from common property and worked on without the permission of potential competitors, another argument against Lockean acquisition is that it lacks legitimacy. The Lockean proviso can be used to refute this issue, but as we have already shown, it greatly reduces the significance of Locke's argument. Furthermore, it has been argued that even if the argument were valid, it would prove too much because it would support parents' ownership of their offspring. Lastly, one may challenge the fundamental premise that everyone has a natural property right to their body. It has been claimed that Locke's premise of a pre-social natural property right to your body and its labour is illogical because ownership is a normative construct that involves claims against others. However, the concept of self-ownership is defended by libertarians like Robert Nozick as a normative theory that is fundamental to the concept of autonomous individuals. The argument that any "patterned" redistribution or taxing of property resulting from lawful acquisition and transfer amounts to a type of forced labour is another way in which they radicalize Locke's view that property acquisition can be viewed as an extension of self-ownership.

1.2.4. Labor and Desert

The desert component of the labour theory may appeal to those who are not persuaded by the notion of expanding self-ownership through labour mixing. The logic of desert arguments deviates from the natural rights account of property rights, but it is absolutely conceivable that property can be obtained based on one's accomplishments. Because ownership is no longer solely justified by a pre-social relationship between the potential owner and her object. Important components of a desert theory of ownership include the social value of individual accomplishments and comparative justice in evaluating individual contributions. There are a lot of problems that need to be fixed even though many writers agree with some form of the desert theory of property. Since it would be illogical to grant property rights as a reward for pointless or even destructive endeavour, some criterion for the social worth of labour must first be established. Any such standards, though, will rely on a notion of the social good and might therefore continue to be contentious. Secondly, criteria for evaluating individual contributions need to be provided, which is also a difficult undertaking. Determining the true value of human contributions based on ethical standards may seem like an almost impossible undertaking in modern societies where production is characterized by intricate divisions of labour. Even in the seemingly straightforward (but blatantly uncommon) scenario of a reclusive scientist who, after years of labour, makes a significant discovery, it can be challenging to balance the worth of his contribution against the efforts of those who raised him, educated him, constructed his laboratory, cleaned his home, cured his liver, or even put his discovery to good use. Additionally, a radical attack on the concept of self-ownership complicates the desert theory.

Individual skills must be viewed as shared resources rather than the rightful possession of each individual, according to John Rawls. Establishing and implementing tenable standards for the allocation of property based on desert may become considerably more challenging if this theory is true, at least to the extent that the distribution of abilities is genuinely the result of a social and natural lottery.

1.2.5. Positivism about Private Property Rights

Other natural rights theories of property will also be impacted by some of the criticisms levelled at Locke's theory. It may seem unlikely that any simply natural relationship could support property rights if they are relational and the subject of valid social claims. Therefore, before a system of rights as such has been institutionalized, thinkers like Thomas Hobbes rejected that anything fair can be asserted about property rights. Everyone will assert a "right" to everything in the natural state. Although David Hume mocks the arguments offered to support current property rights, he contends that we should nonetheless accept their existence and allocation since any attempt to do otherwise would only make matters worse. However, must we accept an unrestricted positivism about property rights if we assume that arguments for natural rights are in fact untenable? Although many authors agree with Hobbes that rights cannot exist in a pre-social state of nature, they nevertheless maintain that certain property rights systems or distributional principles are ethically better than others.

1.2.6. Social Utility

In the debate over property rights, consequentialist explanations centred on social usefulness are common. They all believe that giving people property rights will benefit society as a whole. Aristotle foreshadowed what Garrett Hardin dubbed "the tragedy of the commons": If people may profit privately from shared resources, they won't be financially motivated to conserve them or put effort into the shared source of their income. Privatizing the jointly owned resources could be one way to generate more advantageous motivations, as this could turn out to be harmful for everyone's interests. Private property rights, the market economy, and the effectiveness of economic transactions are typically assumed to be inextricably linked in economic defenses of property rights. In this situation, incentives are just one aspect of the situation. Essentially, a system of private property rights that encompasses complete rights to exchange will result in a system of price-based communication. We may ascertain what is sometimes referred to as the "real social value" of things thanks to this system. Given the circumstances of an ideal market, pricing will represent everyone's unique preferences if resources are privately owned and everyone is free to haggle.

Resources would be distributed in the most effective—Pareto optimal—way possible since the highest bids would represent the strongest preferences and would be accepted first. Therefore, establishing the correct incentives—which would assume that the social good is already known—is not the only benefit of a private property system. Instead, it was impossible to determine the social good without reference to this framework. First of all, this fundamental concept could support a broad theory of property rights: Although it might support the existence of some property rights, it would not necessarily offer guidelines for the first allocation of these rights, in contrast to Locke's theory, for instance. (The "Coase theorem," which is hotly contested in economic theory, states that the distribution of property rights would not even affect the resource allocation that arises from optimal market exchange if transaction costs were absent.) Different principles for their original distribution appear to be compatible with the basic economic rationale for property rights. However, as Ronald Coase and his supporters have shown, social value arguments can also be used to resolve disputes over disputed property rights. They propose to distribute contested property rights to those who will likely earn more, assuming that disputes over property rights should occasionally be resolved by the courts under the realistic premise of positive transaction costs. This proposal will occasionally produce outcomes very different from those based on a traditional Lockean account of ownership, as has become clear in the debate between "Coaseans" and libertarians. Not only Libertarians contend that strict implementation of it may be incompatible with fundamental ideas of justice. However, there is also criticism of the general defense of property rights based on social usefulness. As previously mentioned, it is predicated on intricate philosophical and empirical presumptions, some of which have been contested. Some argue that the "tragedy of the commons" does not always happen and that, in cases where it does, privatizing common goods may not always be the wisest course of action, to name only the most well-known objections. It has been suggested that various ownership rights regimes could serve as the foundation for effective market mechanisms. Many have also argued—most notably Rawls—that any attempt to define the social good in terms of pareto optimality, or the "real social value" of products in terms of their market price, will not be persuasive because these criteria do not take distributive justice into account.

1.2.7. Property Rights and Distributive Justice

If there is already a system of property rights in existence, distributive justice considerations are primarily employed to evaluate different allocations of these rights. Their primary goal is not to resolve the overarching issue of defending the establishment of property rights in general. However, theories of just distribution can take efficiency into account (as in Rawls' difference principle), and these factors may make some of these systems of property rights better than others given empirical knowledge about the expected effects of alternative systems. Accounts of distributive justice, independent of such empirical knowledge, appear to be compatible with a broad range of definitions of property rights; however, they will not be compatible with a libertarian notion of property rights that excludes any such interferences, since the realization of just distributions will likely necessitate frequent interferences with the results of unrestricted market exchange.

1.2.8. Autonomy-Based Arguments

Autonomy is intimately linked to the liberal and libertarian concept of self-ownership. The Kantian notion of humans as ends in themselves is one that Nozick use a lot. However, in contrast to Kantian autonomy, libertarian self-ownership is essentially a negative claim right that safeguards an individual's ability to do with herself whatever she pleases. To enjoy self-ownership, you must have the same authority over your body as a master would have over his slave. As a result, libertarians like Nozick and Walter Block contend that the freedom to sell oneself into slavery is a component of self-ownership. There is conceptual room for autonomy-based defenses of property rights that diverge greatly from libertarian views since this liberty is more closely related to what Kant referred to as "Willku¨rfreiheit" than to his more demanding conception of autonomy.

The relationship between property and autonomous agency is emphasized by Kant and contemporary Kantians. However, there may be other components to this relationship. Access to specific material commodities is obviously necessary for personal autonomy, or the capacity to live an autonomous life at all. Even positive claim ri□hts to property (as defined by human ri□hts) would be justified to the de□ree that property ri□hts are required to ensure this access. Furthermore, as Gewirth argues, property rights may play a key role in establishing a private realm that permits manifestations of individual liberty. In this regard, Kantians can accept Hegel's theories regarding the connection between positive freedom and property.

Kant clarifies that pre-social natural ri□hts arguments cannot be used to determine the exact boundaries of this realm and, consequently, the exact form of property rights. Instead, democratic self-determination must determine them favourably. Lastly, the demands of political autonomy may once more justify property rights that offer equitable chances to engage in the exercise of democratic self-control.

1.3. THE POLITICS OF PROPERTY RIGHTS

A property rights regime can be assessed socially in order to alter it in a desired manner because it is a social construct. Property rights have always garnered significant political attention, so that is nothing new. The dyadic concept of property rights, which pits private against public, is one line of thought in the political debate. The question at hand is whether property rights (namely, land ownership) belong to the person, the family, or a collectivity like the state. At one extreme are those who believe that "property is theft" and that it should be owned jointly, while others view private property ownership as a safeguard against state tyranny. More specifically, the question in the majority of modern Western cultures is: how much and how should the state interfere with the exercise of private property rights? The majority of Western nations concur that private property rights cannot be exercised without restrictions, and even the most ardent supporters of private property rights typically support the state (or other collectivity) owning land and constructing highways on it if doing so allows access to their property. Triadic refers to another line of reasoning in the political debate over property rights regimes. Then, there are three parties instead of just two: the state as an adjudicator in that private-private relationship, one person who can influence another through the exercise of their property rights, and a private individual. In actuality, the majority of property law and custom is civil law: a collectivity governs how (private) legal persons relate to one another and use their property. Dyadic-triadic analysis can be used to understand a large portion of the current discussion surrounding land-use planning and property rights. In order to justify land-use planning by a public entity in terms of welfare economics, Pigou's (1932) arguments are typically cited. Due to some features of the market where private individuals trade property rights, a Paretian welfare optimum will not always be attained. "Market failures" exist. This is viewed as an issue that is fixed when a government agency steps in to protect property rights. Examples include limiting a landowner's ability to use their property as they see fit, taking away their ownership rights, and granting public ownership in order to deliver "public goods." In 1960, Coase made the case that unclear property rights and transaction costs are the primary causes of market failures. Examples of the latter include the right to graze cattle on public lands and the right to clean air. When rights are not fully separated, they become common property, or in the public domain (Barzel, 1898). If these rights are valuable, resources are used to seize them. Rent-seeking and "common-property dissipation" may result from this; Hardin's (1968) "tragedy of the commons" is the most well-known example of this. This argument holds that the issue of market failures can be addressed not only by government intervention to limit property rights (Pigovian measures: additionally, these can exacerbate the situation by causing "public failures;" *see* Wolf, 1979), but also by public measures to facilitate the proper operation of property rights markets, lower transaction costs, and define exclusive property rights. According to economic theory, enhancing the functioning of the private property rights market will lower risks, encourage investment, and result in a more efficient use of resources than would be possible with public land-use planning. The political argument is that a property rights regime should be established where the majority of property rights are owned privately in dyadic terms and where the public administration's job is to regulate private property rights transactions in the methods mentioned above in triadic terms. "Pigovian" measures may be justified if this is not feasible (for instance, because transaction costs are large and cannot be decreased, as is the situation with metropolitan highways). The "property rights model" refers to this method of land-use planning (*see, for example*, Lai, 1997).

1.4. PROPERTY RIGHTS AND THE LIMITS OF PHILOSOPHY

It would seem that a Kantian defense of property rights must balance the Charybdis of leaving too little space for the positive legislature with the Scylla of being overly ambiguous in its definition. This could be interpreted, therefore, as a broad challenge to philosophical conceptions of property rights. How much of those rights are based on enduring philosophical truths? On the other hand, how much do claims regarding the validity of particular property rights regimes rely on potentially speculative empirical hypotheses and presumptions about human motivations that could alter in the future or about relative social preferences? It is obvious that many of the arguments put forth by philosophers rely on empirical presumptions. People normally prefer to own property, for instance, and it has been said that this propensity can only be stopped by a high level of suppression. Nonetheless, there are numerous historical movements—from the Cynics of ancient Greece to Christian or Buddhist monks—where individuals attempted to liberate their spirits by minimising their personal belongings. (Of course, this comment is merely intended to allude to the difficulties of philosophical thinking on social life, not to advocate for a certain style of living.) As a result, certain philosophers, including Rawls and Juergen Habermas, have been very hesitant to make any statements regarding the reasons behind property rights. Others attempt to strengthen their ideas by combining many lines of reasoning. Philosophers may be well advised to exercise extra caution when making generalizations about property rights systems due to the complexity of the issues at hand, the unpredictability of social developments, the contentious nature of pertinent theories in economics and anthropology, and the difficulties associated with pluralism.

1.5. HISTORY

All of our rights, including the right to freedom, are based on property. Property rights "come from culture and community" and define one's ownership of goods in relation to others. A system of laws governing the usage and access to resources must be put into place when a group of people band together to form a community; this determines who is eligible to benefit from those resources. Protecting privacy against trespassing, permitting limited use of shared commodities, promoting productivity by collecting the advantages of ownership, providing stability, preserving peace, distributing limited resources responsibly, and internalizing externalities are all goals of property rights. A property owner is allowed to keep all of their property rights or give some to others while keeping the remainder for themselves. Property rights are commonly referred to as a "bundle of sticks" or a "bundle of rights." The rights to "sell, lease, mortgage, subdivide, grant an easement," utilise the property, and admit or reject individuals are some of the property rights or sticks included in the bundle. Over time, a number of theories about the acquisition and maintenance of property rights have emerged. Locke's Labour Theory is the first. According to natural law, "every man has property in his own person," which includes his hands and body as well as anything he takes from nature and combines with labour. This approach benefits both property owners and society at large by limiting waste and increasing the value of property through labour. All material resources are considered common resources in the second method, Blackstone's occupation theory; the right to property is predicated on first occupancy, wherein a single person asserts complete and exclusive control over a material resource. Lastly, the Demsetz economic model's theory contends that the most efficient result will be achieved with the least amount of waste, externalities, free riders, and overconsumption if property is dispersed and allocated efficiently. The rationale underlying the development of a safe system of property rights is clarified by these theories. Since societies first started to congregate and adopt an agrarian way of life, the incentives outlined by property theories have played a crucial role in the development of organised property rights, which gave people the impetus they needed to make investments in improvements and resources. Over time, these rights were further clarified and are still being established by the legal system today. America's emphasis on property rights was influenced by the English tradition of doing so. Even before America was formally established, colonists benefited from resource ownership and used private contracts more frequently, which promoted the free market economy. When new state governments were established during the Revolutionary War, most of them had clauses demonstrating their commitment to upholding economic liberty and property rights. All people have the "natural, essential, and inherent" right to "acquire, possess, and protect property," according to the constitutions of several states, including Massachusetts, New Hampshire, Pennsylvania, and Virginia. Additionally, following the Revolutionary War, the US Congress passed the Northwest Ordinance of 1787, which also supported the protection of property rights.

The founders of the Constitution took action and "treated private property as the cornerstone of a free society" because they reco□nized the si□nificance of private property rights in the post-Revolutionary age. They held that both economic growth and public safety were facilitated by clearly defined and strictly upheld property rights. "No person shall be deprived of life, liberty, or property without due process of law; nor shall private property be taken for public use, without just compensation," according to the Fifth Amendment of the US Constitution. Additionally, the Due Process Clause of the Constitution's 14th Amendment shields private property against states. In accordance with the framers' goals, state and federal courts have actively defended and upheld private property and contractual ri□hts ever since the Constitution was ratified. During the early years of the United States of America, federal courts supported the framers' property-conscious goals and sought to stop the government from interfering with private property rights. The United States Supreme Court rendered a decision in the si□nificant case of Vanhorn's Lessee v. Dorrance in 1795. The defendable right to acquire property is a fundamental right, according to Justice Patterson, the presiding justice in this case. Property rights were not only a legal construct when they were first established; they were also embodied in each person as being essential to promoting private industry and labour, protectin□ personal property and its benefits, inte□ratin□ into society, and enjoying oneself. The Supreme Court's ruling in Calder v. Bull, which prohibited transferring property rights from one person and giving them to another, strengthened the notion of robust and secure property rights. Property rights were protected by the constitution at the state and federal levels by the turn of the 19th century. Influential jud□es includin□ jud□es John Marshall, Melville Fuller, and Stephen J. Field established a pattern in the US Supreme Court that articulated the notion that safe property rights are the foundation for the protection and safety of all people and rights. Therefore, "protection to one goes with protection to the other; and where either is uncertain, neither can be prosperous nor progressed." The 1800s saw the development and adoption of numerous novel techniques and technological advancements. The courts had to balance an owner's protection in their property with the public's desire for information. By promising that such developments in private property would be shielded from interference, the increasingly important introduction of specific safe□uards durin□ this time, such as patents and copyri□hts, exhibited encouragement in innovation and progress.

During this period, state courts' private property jurisprudence also started to develop, supporting the notion that property rights need to be upheld and safeguarded. The Progressive movement, which attacked property rights, was the vehicle for reform in the United States at the beginning of the 20th century. Nonetheless, in a number of later judgements, the Supreme Court upheld private property rights. A state statute that effectively prohibited newcomers from establishing enterprises in the ice sector was declared unconstitutional by the Supreme Court in the case of New State Ice Co. v. Liebmann. In that instance, the Court ruled that persons have the right to conduct lawful private enterprise without hindrance and that entrepreneurial freedom is equivalent to freedom of speech. Although the pro-private property rights agenda during the New Deal Era in the middle of the 20th century caused some delays, courts largely continued to defend stable and secure property rights. The courts' historical enforcement of private property rights, their desire to be consistent with the framers' intentions, and their desire to support economic growth and technological innovation are probably the main causes of this ongoing administration of private property rights.

1.6. IMPORTANCE OF PROPERTY RIGHTS

Every person's life in the US and around the world depends on their ownership of resources and materials, which is a fundamental right. In a free society, property fulfils two essential purposes. First, through the generation of wealth and prosperity, stable and secure property rights serve as a powerful motivator for economic growth. By declaring that it "takes away from the incitements to industry, by rendering property insecure and unprotected," Justice John Marshall concurred that the preservation of private property rights and contractual freedom were crucial for economic prosperity. Individual prosperity and the market economy will thrive when property is protected by clearly defined and upheld rights, which will result in the transfer of advantages to all citizens in the form of more goods and riches. The second purpose of property is that it has historically been associated with independence, equality, individual liberty, and personal empowerment. "An economic system based on respect for private ownership" "strengthen[s] individual autonomy from government" and makes it easier for power to be distributed throughout society. Because property rights are linked to national wealth and economic growth, they are not only important in the United States today because Americans want to keep up the economic growth and advancement trend, but they are also crucial in developing nations and rural communities worldwide. In order to accomplish important objectives in poverty alleviation, self-governance, gender equality, ecosystem management, environmental and resource sustainability, rural economic growth, and climate change, governments in an increasing number of nations have started to acknowledge the importance of local land and resource ownership. For instance, none of these factors ensures that women will actually have authority over the land, even though they may acquire it through marriage, inheritance, or unofficial networks. The gender gap in property ownership and control is the only major factor contributing to the gender disparity in women's economic well-being, social standing, and empowerment, indicating that the traditional rights of women to land have not been sufficiently acknowledged. Defined and upheld property rights are crucial because the majority of rural people and developing nations rely on ownership rights over land and natural resources for revenue and subsistence.

1.7. POSSESSORY INTERESTS IN PROPERTY

Identifying the objects (items) and subjects (people and groups) of the jural relationships with regard to things in Western legal systems generally is essential to the concept of property. A discussion of possession and ownership, two concepts that have historically been intimately associated in the West, follows. Generally speaking, Anglo-American law is less focused on definitional issues than civil law.

There is little debate in the Anglo-American legal system on whether a particular interest or object should be categorised as property, with the exception of the United States, where doing so immediately grants it constitutional protection. However, Anglo-American law exhibits many of the same traits as civil law. Property interests are thought to be supported by almost all tangible objects; certain intangibles are treated similarly to tangibles, while others are not.

1.7.1. Water

In the West, water and land beneath and adjacent to water are considered differently from other types of property. Western modern law tends to grant the state significant authority over water and territory adjacent to water. In addition, different jurisdictions have rather different regimes (see territorial waters; riparian right). Regarding the taking of water from a stream, whether navigable or not, the United States has a well-developed legal framework. The right to draw water from a stream in the eastern United States is contingent upon who owns the land next to the stream.

The right to drink water in the western portion of the nation is typically contingent upon prior appropriation. Public regulation is becoming more and more prominent in both regions of the nation.

1.7.2. Other Natural Resources

In several Western legal systems, the typical private ownership of other natural resources has been taken away. On the Continent, it is common practice to subject all minerals to state ownership or, at the very least, broad state control. In the past, the crown of England was the exclusive owner of gold, silver, and lead. Private mineral ownership has been the norm in the US, although there is a lot of state control in the name of conservation. Western systems that permit private ownership of minerals alternate between granting them to the landowner and the discoverer, much as private ownership systems for water have tended to split between those that give the water to the person who has it on his land and those that give it to the person who discovered or appropriated it.

1.7.3. The Human Body

The human body, whether alive or dead, is not considered private property in the West. In many legal systems, this fact has caused problems. For instance, if the human body is not property, then what happens when someone donates or sells blood or organs, or disposes of his corpse in a testamentary manner for medical reasons? Although there is particular legislation on this subject in many countries, the conceptual challenge remains unresolved.

1.7.4. Possession of Tangible Things

At least in the West, the idea of possession of a material object predates conscious legal reasoning. According to Roman jurists, possession is a fact that is composed of an object and an intention (animus et corpus). The object was essentially anything that could be physically possessed, with the goal being to claim it as one's own. Seisin, the procedure by which a lord gave his man control of a property, was another social issue that English law had to address. The idea of seisin was also used in English law to refer to material objects that were not lordship property, such as land. Any legal system that bases its property law on the idea of possession would have a system that favors material possessions. Westerners can easily imagine owning nearly anything that can be handled. Possessing an idea such as a right, privilege, or authority is much harder to imagine. People in the West who are not attorneys will claim to own their land or their watches, but they hardly ever claim to have their bank accounts or the authority to sell their land.

1.7.5. Possession of Intangible Things

In the wake of Roman law, civil law has tended to rule out the notion of lawfully possessing something that cannot be touched. In contrast, the idea that someone may have a right, a power, or a privilege is typically more permissive under English and American law. The civil-law systems' greater unwillingness to recognize possessory interests in intangibles has significant implications for how the two systems conceptualize property rights because possession is so essential to both Anglo-American and civil law conceptions of property. Civil law is hesitant to acknowledge property rights in anybody other than the landowner and tends to provide ownership to the landowner. However, Anglo-American law tends to speak of ownership of an interest in land, i.e., of an intangible legal abstraction in a tangible substance, rather than ownership of the land itself, because it recognizes numerous possessory rights in property.

1.7.6. Government-Granted Rights as Property

The 19th and 20th centuries saw a significant expansion in the kinds of intangible rights that governments awarded. The most ancient of them are the exclusive rights granted by governments and international organisations to support and safeguard writers, inventors, producers, and craftspeople. Almost everyone agrees that copyright is a property right, which is the sole authority to forbid the duplication of a piece of writing, art, or music. Copyrights can be freely assigned in the majority of Western systems. Like other types of property, they are typically shielded from state intervention. In most Western nations, trademarks—the government-granted exclusive right to market one's goods with a certain unique sign or symbol denoting its source—and patents—the government-granted right to the exclusive use of an invention—are treated similarly. In the United States, it appears to be obvious that a legislature can donate money to a person or group of people in a way that gives them the right to the grant's property protection. In that case, the grant cannot be revoked without following the proper legal procedures. The award may even be made in a way that makes it impossible to receive it without paying compensation. Court participation in these public-law initiatives has been lower in other Western nations. Therefore, it is probably all the more noteworthy that there has been a trend in the West in recent years to increase the security of at least some types of government subsidies. Generally speaking, there are fewer grounds for removing government funds, and the procedure for doing so has grown more complex. Regarding specific types of agreements between private persons, the same propensity towards property-like treatment is also evident throughout the West. For instance, landlord-tenant law, a traditional area of property law, at least in the descriptive sense, has tended to provide the renter with more security (see Landlord and Tenant below). Western law has also tended to give workers (who do not own property rights in the descriptive sense) more protection. For instance, it requires an employer to provide a reason for firing a long-term employee.

1.7.7. Movable and Immovable Property

The distinction between mobile and immovable tangible things has remained largely consistent, even though the line between tangible and intangible property has become increasingly hazy in Western law and the category of immaterial property appears to be growing. Anglo-American property law started out as a land-related law, as was previously mentioned. Both in the sense that the interest claimed was deemed to be good against the entire world and in the sense that the remedy offered was the recovery of the land or the interest claimed in it, the actions that safeguarded land interests were "real." Movable items, on the other hand, were protected by "personal" actions, both in that the only remedy available was typically monetary damages rather than the specific recovery of the item, and in that one had to claim that the defendant had committed some wrong in order to recover. These two categories of behaviour led to the designation of movable property (like personal belongings) as personal property and immovable property (like a permanent building) as real property (see real and personal property). Modern Anglo-American law has progressively grown to see both types of property as similar, starting with a legislation that created a stark distinction between land interests and all other types of property. However, there are still differences between the two in many jurisdictions that are more the result of historical evolution than of any functional differentiation that exists now. For instance, depending on whether the property being transferred is real or personal, various conveyance types are employed in practically all Anglo-American countries. In many Anglo-American countries, the kinds of interests that can be recognized in the two also differ. The difference between movables and immovables is also acknowledged by contemporary civil law. For instance, the distinction serves as the primary division of property law in Germany, while contemporary Russian law similarly separates property into movables and immovables. However, the civil law has generally minimized the distinction in accordance with Roman law. Certain kinds of usage rights, like a right-of-way or a building privilege, are only acknowledged on land and are often interests that cannot be obtained in a transportable good. The conveyance of movables and the conveyance of land may differ slightly, although not significantly. Land may be subject to longer statutes of limitation or periods of prescription than movables. Overall, nevertheless, the distinctions are not as significant as they are in Anglo-American law.

1.8. SUBJECTS OF PROPERTY RIGHTS

The difference between property-as-commodity (property as wealth) and property-as-propriety (property as the material foundation of the good—i.e., proper—society) has influenced the range of acceptable objects of property, and the agglomerative tendency has had a significant impact on the topic of the subjects of property rights. Anglo-American and civil law both aimed to identify a single legal person as the repository of the enormous array of property rights, privileges, and powers. Although the agglomerative tendency has not been destroyed, historical changes in the law of persons (such as the recognition of more people as having equal status before the law) have produced additional people to whom the tendency could attach. If anything, the tendency has been exacerbated by the fact that contemporary law permits the formation of fictional legal bodies (corporations) without restriction.

1.8.1. Single Individuals

A single human being is the archetypal owner of property in both Anglo-American and civil law. This paradigm has not yet changed despite the fact that a considerably greater portion of wealth is held in corporate or co-ownership arrangements in the West today. Both the ability to interact with property and the ability to hold property are still limited. As a result, the ability of noncitizens to own property is still restricted in many countries. The ability of indigenous non-Western peoples to own property is governed by different laws in several Western nations where they reside. For instance, American Indians living on reservations are subject to such regimes, at least in relation to tribal land. Some limits on noncitizens' use or development of land are limited to agricultural land, while others have a far wider reach in non-Western nations (like Malaysia). Despite having the legal capacity to own property, many persons lack the legal capacity to manage it. Children are acknowledged as being able to own property in most Western legal systems, but they are unable to manage it without their parents' or guardians' permission. Incompetent adults can be stripped of their ability to manage property through procedures included in all Western legal systems. The appointment of a guardian for the incompetent is typically covered by these procedures; the guardian has the authority to manage the property on the incompetent's behalf. In the West, restrictions on the ability of competent adult women to own property and manage it have been virtually eliminated. Although laws in the late 20th and early 21st centuries tended towards equalizing the powers of husband and wife, absolute equality of power to deal with marital property is not the norm in all Western nations. This is because marital property regimes range greatly.

1.8.2. Groups

Western legal systems all permit various types of communal ownership, despite their propensity to view individual ownership as the standard. While not all-inclusive, the categories listed below provide a general idea of the several types of group ownership that could exist.

1.8.2.1. Concurrent Individual Owners

A group of people may each have an undivided ownership interest in anything, according to all Western legal systems. Although it is common for a group of siblings to inherit property from a parent, an individual owner may also sell or donate a piece of property to a group. Tenancy in common and joint tenancy are the two most widely accepted types of co-ownership in Anglo-American nations. Each renter has the privilege and right to own the entire property in both forms. Since everyone has legal custody, they must decide among themselves who will actually have possession if it is physically impossible for them to all own or use the item. One or more of them may ask the court to divide the item among them if they are unable to agree. The court will order the item to be sold and the revenues distributed to the former cotenants if a split in kind is not possible. The authority to convey and succession are two areas where the two types of co-tenancy diverge. If one of the joint tenants passes away under a joint tenancy, the other tenants get his share, sometimes referred to as the "half" or moiety. If one of the tenants in a tenancy in common passes away, his heirs or devisees take over his moiety. In a joint tenancy, the survivorship interest of the cotenants is destroyed if one of the joint tenants transfers his moiety *inter vivos* (e.g., by a living trust). The conveyee is a tenant in common with the other tenants rather than a joint tenant. However, conveyance functions similarly to succession in tenancy in common. The transferring tenant's undivided interest is transferred to the conveyee. Similar to the Anglo-American tenancy in common, civil-law regimes acknowledge a type of co-ownership. In civil law systems, property cannot be held in a way that makes it automatically pass to one's cotenants. Similar to Anglo-American law, French law is adverse to measures to limit co-owners' ability to demand the division of a co-tenancy. However, under German law, cotenants are not allowed to divide the tenancy property (Gesamthandeigentum), even though they are free to alienation their shares. The partnership of coheirs, which endures until the estate of the deceased is settled and distributed, is one of the various types of partnerships that use this type of co-tenancy. Under English common law, each partner had partnership property in their own right. Although they had to report any gains to their partners, each partner's ownership stake was in themselves, not the partnership. In England now, the common-law norm is in effect. However, laws in many U.S. jurisdictions permit partners to hold partnership property in a co-tenancy arrangement called tenancy in partnership, which is quite similar to the Gesamthandeigentum in Germany. Similar to English common law, Roman law addressed joint ownership; however, the rule has generally not persisted in contemporary civil law. The French and other civil-law nations that do not recognize the Gesamthandeigentum as a form of ownership are more likely to acknowledge the partnership's ability to hold property. In these nations, partnerships are therefore regarded as companies for the purposes of property ownership.

1.8.2.2. Marital Owners

In the late twelfth and early 13th centuries, English common law established a system of separate marital property. The husband owned his property, and the woman hers. Tenancy by the entirety, a type of tenancy that is still used in certain American states, was the sole property they possessed jointly that had been conveyed to them jointly. Similar to joint tenancy, tenancy by totality transfers all property to the surviving spouse following the death of the other spouse. In contrast to joint tenancy, it is impossible for one spouse to express their interest in order to undermine the other's right to survivorship. This system underwent a significant transformation in the second half of the 20th century as a result of campaigns for women's equality. Nowadays, a non-barrable statutory elective share is allowed in the majority of American jurisdictions. A standard statute treatment allows a spouse to collect a statutorily fixed portion (usually one-third) of the dead's estate in lieu of any provisions made in the deceased spouse's will. A distinct system of marital property is in place in civil-law jurisdictions and a few U.S. states. Similar to the common-law system, each spouse has their own property, although this only includes any assets they owned before getting married or any gifts or inheritances they acquired while married. Community property includes all movables and, in many civil-law jurisdictions, any property resulting from either spouse's earnings during the marriage. Community property is often split in half following the death of the first spouse, whereas separate property passes to the heirs of the holding spouse. The surviving spouse receives half of it, while the first-death spouse's heirs receive the other half. Unless there is a testamentary decision to the contrary, other community-property jurisdictions give the surviving spouse the first-dying spouse's share of the community. Both the community-property and common-law systems developed during a period when divorce was less frequent than it is now. The current trend in common-law property jurisdictions is to give the judge broad authority to split a divorcing couple's assets without taking into account who is the rightful owner of each item. The inclination in community-property regimes is to divide the community and leave the separate property in the possession of the spouse with title. It is impossible to overstate the significance of marital property for the Western conception of property. The majority of married persons in the West today live under a system of either community property or separate property that is subject to division upon divorce and to a compelled share in the surviving spouse, even though spouses have some ability to alter their marital property arrangements through private agreement. It is reasonable to wonder how much a married Westerner may be considered to own their own property when their spouse owns such a large portion of it.

1.8.2.3. Corporate Owners

False legal entities, such as corporations, sociétés, and Gesellschaften, own the great majority of productive assets in the West. These entities were established under general incorporation statutes that permit them to participate in a wide range of profit-making activities and, often, eleemosynary (related) activities. Even though this development is relatively new, it is so prevalent in modern society that it hardly has to be mentioned. The legal answer to the question of who has the authority to hold, use, and transfer a corporation's property is that the corporation, just like an individual, does. However, people act jointly through corporations; corporations are not individuals. A multitude of interests are concealed by the seeming simplicity of corporate ownership.

1.8.2.4. Community or State Owners

Certain material possessions, like water, air, or wild animals, are taken away from individual ownership in every Western legal system. These things are typically viewed as belonging to the state or the community under contemporary Western law. Additionally, some items that are not taken out of private hands may at any point end up belonging to the state, the community, or another official body. Some of these may be owned by the state in a way that is very similar to things that are in private ownership, such as government office buildings or government-owned businesses; others, such as public parks or highways, may be open to the public generally, at least under specific circumstances. Things that belong to someone else privately may be of interest to the state or the community. A public official acting on behalf of the community or, in certain cases, any impacted community member filing a lawsuit as the community's private attorney general may abate offensive land usage. As environmental concerns have grown, so too have the types and quantity of land uses that are considered objectionable in the West (see below Environmental and historical controls). By using its eminent domain power, the state can obtain ownership of privately held property or individual asset interests. When the state uses this authority, it must compensate the owner, usually as a result of a specific clause in the constitution. Lastly, former communist nations still allow state ownership, just like the West. For instance, Russia recognizes both private and state ownership. State ownership, whether it be at the federal, state, or local level, is referred to as public ownership.

1.8.3. Unitary and Nonunitary Concepts of Ownership

The idea of ownership is interpreted unitarily in the civil-law system. The "triad of ownership," which includes the owner's rights to hold, use, and dispose of an item, is a term frequently used by civilians, especially those in post-communist legal systems like Russia's. In this way, the civilian understanding of ownership is likely more similar to that of regular non-lawyers than it is to the Anglo-American understanding. Ownership is not viewed as a singular term in the common-law community. Anglo-American solicitors, on the other hand, view ownership as a "bundle of rights." This popular metaphor alludes to two aspects of ownership. The first is the potential for ownership to be dispersed. The fact that the object of ownership, or the "thing," can be owned by multiple people draws attention to the particular restricted rights that each of the different co-owners has over the item. Similar to the first feature, the second one highlights the many rights that different people may have to the item at the same time. Therefore, although while the person who is commonly referred to as "the owner" may have the right to own, use, and dispose of the item all at once, in reality, someone else may hold one or more of these rights. For instance, although if a home's occupant has the sole right of occupation, the person who owns the mortgage is legally regarded as the home's owner in several U.S. states. Ownership and possession are closely linked in both legal traditions. Nonetheless, Anglo-American law permits the right of ownership to be divided across time, allowing one individual to possess during their lifetime and another to possess after that (see Temporal divisions below). On the other hand, civil-law regimes hardly ever permit such a temporal division of the right of possession. Civil law is more likely to conclude that the individual who has the right to possess for his lifetime merely has a "usufructuary" (right-to-use) interest rather than ownership in the example above. Since the right to possess and ownership are linked in Western systems, it is possible that property ownership will change if the right to possess and actual possession are kept apart for an extended period of time. If person A leaves a piece of land he owns and person B acts as the true owner would, person B may take possession of the land under a body of law called as "adverse possession." Person A must take action to regain his ownership from Person B within the time frame specified by the statute of limitations, even when that second possession is unlawful in his opinion. The statute of limitations for land recovery actions is typically 10 or 20 years in the majority of Anglo-American jurisdictions. However, person A's action will be prohibited if he does not take action within the allotted time. Who then owns the land, one would wonder? The peaceful owner of land has the right to keep it in the majority of Anglo-American countries, with the exception of those who can demonstrate a stronger claim. However, person A's claim is no more superior than that of the peaceful possessor if the statute of limitations prevents him from obtaining possession. As a result, the individual who actually possesses land during the statute of limitations gains the right to possession against everyone else, even the rightful owner whose claim is now prohibited. Thus, over time, this unfavorable possessor turns into the rightful owner. Although the terminology is different in civil-law nations, the outcomes are comparable. The possessor is stated to obtain title through a procedure called prescription as time passes (a little longer than in the Anglo-American systems).

1.8.4. Property Rights Divisions of Ownership

1.8.4.1. Spatial Divisions

Property can be divided along spatial lines by the owner in all Western legal systems. For instance, when the resulting parcel of land lacks access to a public right-of-way, such divides might be foolish. Public regulation may stop the division in the case of land. A vertical divide instead of a horizontal one presents a somewhat different set of issues. Such vertical divisions are generally permitted by Anglo-American law, allowing one individual to control the land's surface, another the mineral strata beneath it, and a third the air rights. Due to the mediaeval dictum "Cuius est solum eius est usque ad coelum et usque ad inferos" (Whoever owns the soil owns all the way to heaven and all the way to hell), the civil-law systems have had some trouble with this kind of ownership split. For instance, ownership of an apartment on the 30th floor of a building is now conceivable in both systems thanks to contemporary legislation. Owning a condominium is more complicated because the owner not only owns the space inside his home or flat, but also has the rights and privileges to use utilities and communal facilities. By allowing each cooperator to hold a part in a firm, cooperative ownership circumvents this complication. In exchange, the corporation retains title to all of the property while permitting the cooperators to occupy their living units.

1.8.4.2. Temporal Divisions

The quantity and complexity of temporal divisions of ownership permitted by Anglo-American law are well known. When it became impossible to have legal ownership divided temporally other than between landlord and tenant in 1925, the English legislation on the subject was significantly simplified. However, English law still permits complex temporal divisions of beneficial interests in trusts, which permits a temporal division of equitable ownership but not legal ownership. Although it is increasingly done through trust, temporal division of the legal ownership of property is still conceivable in many of the surviving Anglo-American countries.

1.8.4.3. Life Estate and Remainder

The life estate and the remainder in fee, one of the potential temporal divisions of ownership under Anglo-American law, has already been examined. The life tenant in such a contract is entitled to the land for the duration of his natural life. He is allowed to use the property, but he is not allowed to waste it or reduce its capital value. He may express his curiosity, but it will only be restricted by his life and nothing more. As a result, his conveyee is given an estate that is restricted by the conveyor's life (estate pur autre vie). Common-law life estates include curtesy and dower. After the life tenant passes away, the remainderman is entitled to possession. He can't use the land until the life tenant passes away, but if the life tenant wastes it, he can sue them. If the remainderman passes away before the life tenant, his interest will transfer to his heirs or devisees because it is an interest in fee. Subject to the life estate, the remainderman may also express his interest *inter vivos*. A reversion occurs when the original owner in fee transfers a life estate to another person while keeping the remaining portion of the property for himself. Reversions share many traits with remainders for the most part. Anglo-American law allows for several variants on the fundamental life estate and remainder pattern. Example: "to my wife, Edith, for her life, remainder to my son George for his life, remainder after George's death to George's children." These are examples of consecutive life estates.

1.8.4.4. Contingent Interests

In Anglo-American law, it is not only conceivable to establish successive land interests, but also interests that are subject to specified conditions. Therefore, in the aforementioned case, the donor might condition the remaining amount in George on George reaching a particular age, let's say 21, at the time of the prior life tenant's passing. In the majority of Anglo-American jurisdictions, it is not only conceivable to make future interests contingent, but it is also possible to make current interests contingent. As a result, it is feasible to grant a fee interest with the condition that the land be used for educational purposes and to stipulate that the interest would be forfeited if it is not (fee simple determinable, fee simple subject to a condition posterior).

1.8.4.5. Civil Law

Civil law allows for some, but not all, of the arrangements mentioned above. In this sense, the main difference between Anglo-American and civil law is that the former typically does not view such agreements as entailing ownership splits. Therefore, the usufruct—the civil law provision that most nearly resembles the Anglo-American law's life estate—is viewed as a right in another person's property rather than as a type of ownership (jus in re aliena). In civil law, the usufructuary typically has possessory remedies against third-party wrongdoers, even if he typically does not have the right to ownership. Overall, despite the significant conceptual distinctions between the two systems, one can wonder how different the usufructuary's practical position is from that of the life tenant under Anglo-American law. The disparities between the two systems are not as significant as they may appear, even when it comes to conditional presents. Indeed, the fundamental tenet of civil law is that gifts cannot be subject to conditions. The donor must make a full or partial donation. Nonetheless, the mediaeval Roman law of fideicommissary substitutions—the Roman fideicommission that allowed testators to circumvent some limitations on legacies that existed under the formulary system of Roman law—is the source of some exceptions in civil law. The regulations are intricate and differ depending on the jurisdiction. For instance, in the French system, a will can be made granting property to one's children and mandating that they transfer it to their offspring. According to German law, a succession can be appointed as long as it takes place within 30 years of the testator's passing. The fee simple with a forfeiture clause has no counterpart in civil law. Therefore, a donation with the requirement that the land be used for educational purposes cannot be made under civil law, while there are ways to accomplish the same goals under civil law, at least temporarily.

1.8.4.6. Landlord and Tenant

Present possessory rights under Anglo-American law that are smaller than the fee are not always restricted to the holder's lifetime; they can also be restricted to a certain number of years or to a renewal term. The relationship between a landlord and tenant is established by such a transaction. For any given term, whether one month, one year, five years, or 99 years, the tenant may have a possessory interest. Additionally, the tenant could have an interest for a certain duration that automatically renews unless either the landlord or the tenant gives notice within a predetermined window of time before the term ends (periodic tenancy). Tenancies can therefore be set up, for instance, from week to week, month to month, or by year. Another option is a tenancy at will, which is based solely on the wishes of the landlord and renter and has no set duration. The tenancy may be terminated at any moment by either the landlord or the tenant giving notice to the other. (Tenancies at will are more equivalent to periodic tenancies in many jurisdictions where they are subject to statutory regulations regarding the notice period to terminate.) The establishment of landlord-tenant relationships is also permitted by civil law. The majority of the landlord-tenant arrangements recognized by Anglo-American law can be created by private agreement, even though the kinds of tenancies recognized by Anglo-American law do not exist in civil law. The way that the landlord-tenant relationship is conceived under civil law is different. Like Roman law, current civil law grants the landlord the right of possession, not the renter. The practical ramifications of this conceptual distinction are minimal because the landlord is contractually obligated to permit the renter to acquire. The scenario in which the landlord sells his stake in the land to a third party is the primary area where the two systems diverge. The renter has an enforceable possessory interest against his new landlord under Anglo-American law. The tenant's civil law remedy is to sue his former landlord. Recent laws in civil-law states that let renters to sue third parties who really interfere with their possession have even reduced this disparity.

The laws governing landlord-tenant relationships underwent significant alterations in the second half of the 20th century throughout the West. The law pertaining to residential tenancies, namely tenancies in urban apartments, was most significantly impacted by these revisions. The laws pertaining to agricultural tenancies were also significantly altered in some jurisdictions. Commercial tenancy legislation was largely left up to private arrangement. As a result, commercial leases still follow conventional formats, with the parties negotiating the terms. The courts in the United States were the ones that started the modifications to the residential landlord-tenant law. The courts' actions have been confirmed by legislation, which has often gone beyond them. Legislators in the rest of the West have taken the initiative, but the outcomes have been very similar. The 1960s and early 1970s saw changes in U.S. landlord-tenant laws as the nation began to pay attention to the deteriorating housing conditions for the urban poor, who had been left behind as the middle class relocated to newly built suburban housing. Poor people's urban housing was often kept at levels well below what municipal regulations demanded, although these standards were rarely enforced. American courts eventually came to read the housing code's provisions into the lease due to this circumstance and significant legislative inertia. For instance, Javins v. First Nat'l Realty Co. (1970) mandates that a residential lease include an unwaivable warranty of habitability, which requires the landlord to keep the property in accordance with the local housing law. The landlord cannot dismiss a tenant for failing to pay rent, but the tenant may withhold payments if the landlord fails to keep the property to this standard. In the 1970s, laws tended to support Javin's findings. Tenants in residential buildings now have greater tenure security thanks to other laws and rulings. Rent control was re-established in several metropolitan jurisdictions that had previously abandoned it. The tendency is towards convergence, even though the American and European regulatory frameworks for the residential landlord-tenant relationship still diver□e si□nificantly. The fundamental conditions of the relationship are increasingly set by law everywhere; the landlord is now seen more as the supplier of a public utility than as the owner, and the tenant's stake in his home is now more secure. Over time, public opinion has fluctuated between support and opposition to rent regulation.

1.8.4.7. Trusts

The ability to manage property and the right to benefit from it are two further conceivable divisions of ownership recognized by Anglo-American law. In Anglo-American law, this division—known as the trust—has significant practical implications. Family settlements and charitable giving are two of the many situations in which the trust device is employed. It has substantially supplanted the legal life estate and remainder in the context of family settlements. The separation of legal and equitable ownership is central to the concept of the trust. Separate English courts served as the genesis for this split. Common law courts acknowledged and upheld legal ownership, while equity courts acknowledged and upheld equitable ownership. However, even when the law and equity courts merged, the conceptual distinction between the two forms of ownership persisted. Therefore, although though legal and equitable interests are typically upheld by the same courts nowadays, they are nonetheless conceptually separate. The fundamental difference between equitable and legal ownership is rather straightforward. The right to possession, the privilege of use, and the authority to transfer these rights and privileges belong to the property's legitimate owner (trustee). The beneficial owner (beneficiary, cestui que trust) is the only person who does not believe that the trustee is the true owner of the property. The beneficiary receives all of the property's benefits, as agreed upon by the trustee and the beneficiary. The beneficial owner owes the trustee a fiduciary duty to use his legal rights, privileges, and powers in a way that benefits the beneficiary rather than himself. If the trustee doesn't, the courts will force him to give the beneficiary what he has earned for himself. In severe circumstances, they may even take away the trustee's legal ownership and replace it with someone else. Legal and beneficial ownership are typically separated by a specific trust document. The creator of the trust, known as the settlor, will transfer assets to the trustee, who may be a person or a business, such a bank or trust firm, and give the trustee instructions on how to handle and administer the assets for the benefit of one or more trust beneficiaries. Trust instruments are often very complex. Their provisions may cover both the beneficiaries' and trustees' succession. They might grant the trustee a great deal of latitude in overseeing the assets and allocating benefits to the recipients. As with a spendthrift, or protective, trust, the beneficiary's interest may be shielded against his creditors' claims in almost every country. There are laws in several U.S. states that allow people to set up spendthrift trusts for themselves. These so-called "asset protection trusts" shield the owner's beneficial ownership interests—such as the lifelong right to collect all income from the trust assets—from his creditors while preserving the owner's ability to enjoy all the advantages of ownership.

Using a trust, as opposed to the formal life-estate-remainder arrangement, efficiently keeps asset administration and pleasure apart. The trust arran□ement sometimes enables assets to be managed more professionally than they would be by part or all of the beneficiaries by placin□them in the hands of professional or semiprofessional administrators. Furthermore, the agreement does not bind the assets themselves. The trustee can grant good title to the assets if it makes sense to sell a piece of land or switch an investment portfolio from bonds to equities. Furthermore, by employin□a trust, the beneficial interest in property can be passed down from one generation to the next without going through probate, which is a cumbersome and time-consuming procedure in many Anglo-American jurisdictions. Lastly, there are instances where using the trust results in tax savings. The trust has no exact counterpart in civil law. The waqf is an institution under Islamic law that resembles the Anglo-American trust in several ways. An institution similar to the trust has been established in certain contemporary civil-law systems, but typically this has been accomplished by incorporating trust concepts from the Anglo-American system rather than by creating original concepts. The majority of the purposes for which the Anglo-American trust is used are accomplished in various ways under civil law. For instance, the civil-law foundation closely resembles the benevolent trust found in Anglo-American law. Among the aforementioned uses of private express trusts, solicitors on the Continent obtain expert asset management by entrusting their assets to managers who receive compensation for their work. It is less essential to have a trustee who can grant sound title to the entire property under civil law as there are less potential outstandin□interests in a specific piece of property than under Anglo-American law. Since civil-law probate processes are typically si□nificantly less onerous than An□lo-American ones, avoidin□probate is rarely a problem on the Continent. Consequently, it turns out that some of the needs that the Anglo-American trust serves are not needs in the civil-law systems due to structural differences between the systems; the remaining needs are met by other techniques. This is the case in many other areas of comparative Western law.

1.8.4.8. Security Interests in Property

In all Western legal systems, there is an additional split of ownership rights, privileges, and powers that takes place when an owner uses his property as collateral for a loan or other commitment. Despite significant differences in terminology and conceptualization of property utilised in a secured transaction, there aren't many practical differences between the Anglo-American and civil-law systems in this area. Both systems acknowledge agreements between a debtor and a creditor whereby the creditor is ostensibly granted ownership of the item, but the creditor's handling authority is restricted so that the item will return to the debtor as long as the debtor fulfils his obligations. Both systems also acknowledge agreements where the creditor has enough rights against the debtor to be safe in the event that the debtor defaults on his commitment, but does not have an ownership interest in the property. The most intricate and historically significant security mechanisms in both systems relate to land: the common law mortgage and the civil law hypothec (pledge). Under a common law mortgage, the debtor (mortgagor) transferred his land to the creditor (mortgagee), with the understanding that the land would automatically return to the debtor if the debtor fulfilled his end of the bargain by a specific date. However, the debtor continued to own the land, and the practice of letting the debtor keep the land turned into a duty on the part of the creditor to permit the debtor to do so, and ultimately a right in the debtor to do so, provided that the debtor was not in default on the debt. The creditor's right to possess was completed in the event of the debtor's default, allowing him to access the property, use it for his own purposes, and sell it as he saw fit. The interests of the debtor were terminated. The debtor's viewpoint was taken by the equity courts. Even if the debtor had fallen behind on the payments, equity initially granted him the ability to redeem the property by making the required payment. Creditors were compelled to file an equity action to foreclose the debtor's equity of redemption in order to sell the property. Equity granted the debtor the right to the sale proceeds as a condition of foreclosure, provided that the selling proceeds exceeded the outstanding debt. The debtor's right to redeem even after the creditor had foreclosed was increased by laws in the 19th century. Lastly, laws in some jurisdictions mandated that the creditor sell the property after foreclosing, and in some of these areas, a public authority had to conduct the sale.

Since he did not own the property, the debtor was unable to give third parties legal title to it under common law. (However, he may express his equity of redemption.) This meant that even while the mortgagor appeared to be the property's owner since he was in possession of it and could typically provide proof that a former owner had transferred it to him, a buyer acting in good faith might wind up with nothing. The majority of Anglo-American jurisdictions have public offices where mort□a□e transactions can be recorded or re□istered in order to safeguard third-party purchasers (see Registration and Recordation below). Priority in title under common law, or between subsequent grantees, was established by the dates of the corresponding conveyances. Because A was first in line, A wins out over B if O □ave A land and then □ave B the same land. Today, B is protected against A if A takes land from O without recording the deed and B later buys the land from O without actually knowing about the O-A deed. Even if the mortgage deed states that the creditor is the owner, the mortgagor's interests appear more like those of an owner than those of the creditor. Subject to all of the aforementioned requirements, several jurisdictions continue to hold that the creditor is the owner. There aren't many real-world differences between the two kinds of jurisdictions. Despite having rather different starting points, the civil-law systems have mostly reached the same conclusion. Until he defaults, the debtor is entitled to possession and the opportunity of using the property. Depending on the jurisdiction, the creditor may either seize the property or compel a sale if he defaults. Everywhere, the debtor's stake in the selling proceeds beyond the remaining loan balance is safeguarded. After default, the debtor may also be granted a grace period in certain jurisdictions during which he might redeem the property. Security interest registration is almost widespread. Even if the property is transferred to a good-faith buyer who isn't actually aware of the security interest, the creditor's interest will still be there if the interest is registered. The history of security interests in movables is a little different. With the help of legislation in the 19th and 20th centuries, the equity courts under the Anglo-American system played a major role in developing security interests in personal property. As a result, a somewhat complicated area of what is typically referred to as commercial law (see business transaction) is created. To put it simply, there are two possible arrangements that are similar to a mortgage: either the creditor takes possession of the property with the debtor's right to redeem it by paying the debt (pledge or pawn), or the debtor keeps possession of the property subject to a security interest in the creditor (chattel mortgage or conditional sale). In some countries, most notably England, the debtor will lease the property from the creditor, who is typically the seller as well. After the payments are received, the debtor's title will become absolute (hire purchase). Uniform regulation that addresses all personal property security arrangements under a sin□le topic has si□nificantly minimized the disparities between the various types in the United States. (See installment credit as well.) Historically, the primary security mechanism for movables on the Continent was the pledge or pawn (pignus). The creditor had the right to possess the movable under this mechanism, even thou□h actual possession mi□ht not be. Merchant financin□ methods are governed by different rules of commercial law, and they typically resemble conditional sales or Anglo-American chattel mortgages. Numerous tools have been developed by modern consumer-credit law; some are more akin to the English hire purchase, while others are adaptations from the civil law of pledge.

Chapter 2

Property Rights Regimes

Contents

OVERVIEW

Through technology, especially its institutions and the norms and regulations that govern behaviour, society engages with nature. An essential component of effective and sustainable natural resource management is the creation of institutions or property rights regimes that are in harmony with the processes of the underlying ecosystem. There are significant effects on income distribution, efficiency, the economy, and the environment when property rights are distributed for different uses of a natural resource. This section summarizes the current understanding of how communities and natural resources interact under various property rights regimes and how that interaction may impact resource system performance. This chapter's theoretical findings heavily reference recent advances in the property rights field in economics. Given the technical nature of this material and to make the presentation more accessible to non-specialists, the majority of the supporting technical details and proof have been left out, and the reader is directed to the relevant literature. According to economic theory, a stock of natural resources is a productive asset that must be properly maintained and managed to produce the maximum output. It depreciates if it is misused. Because it affects the motivations of individuals who exercise these rights or work with the resources, the property rights system is crucial to the effective management of the resource. The system of property rights may be established historically and upheld as a social norm, or it may be enforced by the government or local authorities. The regime governing property rights may alter in response to shifts in a natural resource's basic "parameters," such as its yield, intensity, and predictability. For example, if resource consumption becomes less intensive, a territorial management system may become a looser, communal arrangement. If local common property institutions stop operating, for example, because of government intervention, it may develop into an open-access system. In his research on the hunting rights system of the native Indians in James Bay, sub-arctic Canada, where the ownership regime initially incorporated community territories, Berkes (1989) offers a specific example. Following the growth of the fur trade, a system of family-controlled areas developed as the resource's (mostly beaver) consumption increased. These local institutions were undermined and eventually eliminated as a result of competition among fur companies and the influx of outsiders. The wide range of property rights regimes, particularly in historically based resource management systems, is a remarkable aspect of the extensive empirical literature examining the relationship between property rights regimes and economic performance. For instance, there is a wide range of arrangements that vary from island group to island group in Oceania's reef and lagoon tenure systems. In contrast to this variation, Western resource management guidelines are more or less the same.

The goal should be to increase the diversity of institutional arrangements and the flexibility of resource management, not decrease it, in order to accommodate environmental and social perturbations. Therefore, enhancing our knowledge of how property rights regimes function in connection to natural resources and common qualities is a crucial study goal.

Efficiency is the criterion used in this section to evaluate the "success" of managing natural resources. Efficiency and sustainability are two different ideas. Even if a resource management pattern is effective, it might not meet sustainability standards. If it is possible to improve the lives of many people without negatively affecting the rest of the community by moving from resource use Y to resource use X, then resource use X is roughly more efficient than resource use Y. This concept is comparable to stating that the community's overall economic surplus under resource use X is greater than under resource use Y, with specific technological assumptions." Many effective uses of natural resources frequently coexist, and the decision entails unavoidable welfare trade-offs between both current and future generations as well as between social groupings (landlords and peasants, rural residents and industrial groups). To resolve these trade-offs, a well-defined and operationalized idea of sustainability could be helpful. Efficiency, however, comes first. Before addressing how the cake should be divided between the current and future generations, as well as between the rich and the poor within the current generation, we first address the issue of maximizing the cake's size given the components. We must keep concerns about efficiency and income/welfare distribution as distinct as feasible. The Hardin (1968) "tragedy of commons," a widely held belief, does not accurately describe what actually occurs in many everyday situations. Instead, a "bucket brigade" metaphor is suggested by a large portion of popular literature. Similar to how a bucket brigade is formed to put out a fire in a rural community, a group of individuals will organize themselves and distribute property rights when faced with a resource management crisis. The tragedy of the commons only occurs when a shock—such as technical advancements, governmental interference, or abrupt changes in the resource's economic yield—destabilizes the relationship between the management regime and the resource users. When stabilizing feedback is available and when it may worsen are insights from the theory of recurrent games, which are summed up here.

2.1. DEFINITION OF PROPERTY RIGHTS REGIMES

Why should the allocation of property rights related to the different uses of a resource or asset have an impact on its effective management? What does possessing a natural resource entail? like a tract of land or a waterway? It is very helpful to think of a resource as a collection of rights rather than just an object, each of which can be independently held by or transferred to various people or organisations. For example, what rights are attached to a 4ha maple forest? A steady supply of commodities and services is promised by this composite resource. It can be sold or leased, utilised consumptively (e.g., for actual construction), or physically separated into areas. A natural resource's ownership can be divided into different categories by examining the rights associated with its uses and services. For instance, the right to collect and sell certain foods like mushrooms or maple syrup, the right to graze in the forest for 20 years, and the right to receive money from tourism and recreational activities. Due to potential conflicts, some of these services and related rights might not be available at the same time. Some are exercised simultaneously, as in the case of the Huerta system in Valencia, Spain, which prohibits the separation of water rights from the land being used, or by a multinational paper firm that manages its forest resources for both timber and wildlife. A coordinated exercise of the different rights connected to a resource may be required by externalities, either positive or negative. Because of uncertainty, ignorance, the limitations of the human intellect, and the time required to come up with such a list, the list of rights to all possible goods and services that can be produced from a natural resource, as well as the conditions under which the corresponding user rights can be exercised, can be extended indefinitely, but it will always remain incomplete. Furthermore, property rights cannot be transferred or exercised unless they are enforceable. These facts have significant ramifications for the ownership structure: all rights except those that are legally restricted or granted to other parties through contracts will belong to one party (the government, an individual, or a collection of individuals). In other words, according to the new idea put forth by Grossman, Hart, and Moore (*see* Grossman and Hart, 1986 and Hart and Moore, 1990), there will be an owner of residual rights who is simply referred to as "the owner." Who this owner should be and what rights should be effectively granted to whom through contractual ties are determined by the criterion of efficient administration of the resource. The definition of ownership given above is used here, and the procedures for assigning these rights to natural resources as property rights regimes are mentioned. Property rights regimes are likely to be varied and complex due to the wide range of services that natural resources can offer. This chapter makes a distinction between government control of the resource base, private ownership of the resource base, and open access regime.

2.1.1. Open Access Regime

Land and related natural resource property regimes must be viewed as a component of a society's broader institutional framework (Bromley, 1999).

Stated differently, property regimes cannot be viewed as distinct from the society in which development issues are addressed since they are a social construct. To define the property regime over land and associated natural resources, or whether it is what we would refer to as private (person) property, state (royal) property, common property, or open access (none) property, sets of institutional arrangements are therefore continuously constructed. In this pretty extreme situation, property rights are just not granted. Since no one has rights, everyone can use the resource.

Only when the resource's "shadow price" is zero—that is, when there is no social cost associated with consuming or extracting a marginal unit of the resource—can an open access regime be considered effective. This necessitates that the resource pool be quite substantial in relation to the community's size and needs, which is undoubtedly a speculative situation given the state of the globe today. The first person or group of people to appropriate resources in an open-access resource situation automatically becomes the resource's owner (Bromley, 1999). Under open-access conditions, ownership of resources results via capture rather than prior state legitimization.

People can utilise a resource freely under open access without considering the consequences for other people. In these situations, we say that a person has the right to use the resource but not the authority to stop others from doing so. The rule of capture forces all users to grab as much as they can, as fast as they can, making an open-access resource a free-for-all. For instance, open-access land tenure happens when traditional or modern land use is not effectively regulated (Cleaver and Schreiber, 1992).

This makes it possible for farmers and others to use the land and its resources in ways that are not sustainable. Additionally, land occupation and use in open-access are transient, and farmers are not incentivized to engage in farm intensification. Additionally, because the property is not theirs, open-access lessens the incentive for farmers to preserve it. Situations involving open-access resources typically arise from the lack or disintegration of a management and authority structure whose main objective was to establish and uphold a set of rules for participants' conduct towards the natural resource.

When the first party to effect capture has access to valuable natural resources, it is either because those resources have never been included in a regulated social system before or because institutional failures have made them open access resources, undermining previous collective or private property/ management regimes (Bromley and Cernea, 1989).

2.1.1.1. Depletion of Resources and Open Access

Resource depletion is always the outcome of open-access arrangements, in which all possible users have total control over how they utilise a resource and no one has the legal authority to stop others from using it. Economic theory backs this up, demonstrating unequivocally that unchecked free access would lead to high rates of exploitation.

As stated by Usually, two variables lead to this outcome. Agents will first enter the resource market to take a portion of the earnin□s or "rents" that are present. Second, even though this would reduce the total amount of rent available in the future, individuals who are already involved in the activity have an incentive to strive to capture more profit even if others are prohibited from participating, say, harvesting and using a resource. Open access can frequently result in the total depletion of a resource if everyone behaves sensibly. Because a competitive □roup will view any financial benefits from resource protection as "excess profits," they will consistently overuse an open-access resource. It is always reasonable to assume that group will react to those rewards by increasing pressure on the resource, which is consistent with Marshallian theory of long-run competitive equilibrium.

Therefore, everyone considerin□ conservation is aware that any benefits will be offset by greater pressure from other resource users to deplete resources. A straightforward graphic model can be used to show how rent is distributed when a scarce resource is freely accessible. Examine Figure 2.1, which shows a fixed common pool resource (such as a □razin□pasture) that needs labour to harvest its resource flows. VMP shows the mar□inal product, while VAP is the value-of-average-product curve that results from labour input. The (exogenous) market wage in alternative activity determines the opportunity cost of applying labour to the natural resource. The first peasant will assi□n 11 units of labour to resource exploitation in order to capture rent equal to the area DEFB since they will equate the pay rate with the value of the marginal product (*see* Figure 2.1). More peasants enter the grazing area since there are no exclusive rights to the resource, which lowers marginal production. They will increase labour until the value of the average product equals the wage rate (12), at which point the value of the marginal product may even be negative, because they do not have to bear the entire loss in marginal productivity; some of it will be externalized to other peasants. The increased harvesting effort will dissipate all rents in the absence of institutions to regulate resource use (prisoners' dilemma). Triangle CFG, the region where MC (marginal cost) is located above VMP, illustrates the inefficiency brou□ht about by the excessive effort put into the fixed resource.

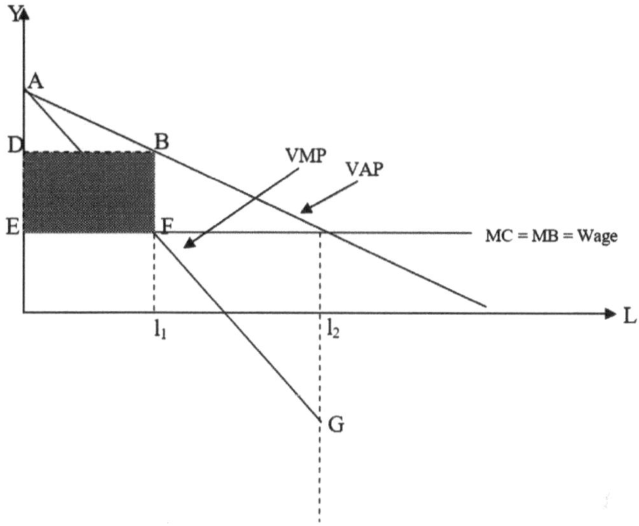

Figure 2.1. *Rent dissipation in the common pool.*

Source: Wachter (1992).

2.1.1.2. Improving Open-access Management

There will be no long-term benefit to efforts to enhance open-access resources without changing their open-access status. Depending on the circumstances, a policy decision must be made regarding whether to convert open-access land to state property, private individual property, or common property. Whichever regime is selected, it will take effort to create a new set of regulations that must be communicated to all relevant parties, as well as new management styles and social authority structures to ensure adherence to those regulations (Bromley and Cernea, 1989). Common property should be the ideal property regime for managing open-access resources because it has been utilised by many societies throughout history to manage natural resources sustainably, according to several writers (Ciriacy-Wantrup and Bishop, 1975, for example). The prevailing argument has been that people who have access to common properties do not always choose activities that deplete natural resources, despite the constant risk of over-use because any one user will feel that the benefits are internalized while the costs of resource depletion are shared by all users. This is due to the fact that they are all aware that allowing their livestock to overgraze at one point will likely discourage others from doing the same in the future. Each person will put up with some short-term sacrifice for the sake of conservation that benefits the group to which he belongs, to the extent that his resource use decisions serve as a signal affecting others' resource use decisions in the future. Additional justifications for turning open-access resources into common property include the fact that, in many cultures, local group norm compliance serves as a powerful deterrent to antisocial behaviour (Bromley and Cernea, 1989). Additionally, a strong common property regime contains a system of financial and non-financial incentives that promote adherence to established institutions and customs. Studies have shown that some environmental issues are less severe when open-access resources are privately owned and managed, despite the fact that bringing them under a regime of individualized private property has been argued to have the potential to conflict with prevalent socio-cultural values in developing countries (Bromley and Cernea, 1989). For instance, if the owners of the land have a clear legal title, the land is less likely to be overused. In a similar vein, those who cultivate property and have secure rights to it are more inclined to manage it over the long term. More secure tenure, access to formal credit, and land investment are all positively correlated, according to one of the few thorough studies on the relationship between increased security and better land management carried out in Thailand. Because private property regimes have the social and legal authority to keep out excess population and successfully fend against undesirable incursions using the state's power, they appear to be stable and adaptable (Bromley and Cernea, 1989).

Open-access resources should not be nationalized in order to be managed more appropriately for the following reasons:

- The majority of governments have little power to enforce laws pertaining to resources owned by the public;
- Frequently, national governments have established an ownership structure without the corresponding ability to regulate the behaviour of individual users.
- Promoting nationalization entails the risk of denying livelihood to significant segments of the populace without fulfilling the anticipated promise of improved resource management; and
- Resources managed by government agencies have continued to deteriorate as a result of the appearance of management brought about by the creation of government agencies and the appearance of cogent policy created by the issuance of decrees that forbid access to state property (Bromley and Cernea, 1989; Bromley, 1990).

Governments must, however, build the institutional and human capacity necessary for managing open-access resources and for keeping an eye on how those made available for exploitation are being used, especially in cases where government ownership of these resources is unavoidable due to their particularly valuable environmental characteristics. This is necessary to guarantee that those intended for protection are indeed safeguarded and that those made accessible for exploitation are exploited in a sustainable and productive way, according to Cleaver (1993). Another important necessity is that the government rectify its current indifferent enforcement procedures and provide enough people and incentives to oversee and administer the territory it has appropriated for itself (Bromley and Cernea, 1989). A system where the government owns the resources but grants people and organisations long-term, legally binding user rights could also be a suitable choice. Long-term, secure user rights encourage conservation (Cleaver, 1993). For instance, in the Philippines' Central Visayas region, the majority of the land is owned by the states, and the tenants' land tenure is highly precarious. This resulted in low-cost, short-term production methods and the depletion of natural resources. In order to mitigate these adverse consequences, the stewardship program offered leases to public land users that gave them 25-year use rights with the possibility of extension to 50 years, provided that they adhered to a resource management plan (Wachter, 1992).

2.1.2. Common Property Regimes

Common property regimes are resource management strategies based on a set of mutually agreed-upon rights and guidelines for the sustainable and mutually beneficial use of shared resources. A resource under the management and control of a group of users is called a communal good. A well-defined resource that the group uses and maintains, a recognized set of users, and institutional arrangements (rules and rights) for the use of the resource make up a common property regime. It has been demonstrated that common property regimes, which are collective resource management systems, emerge when a group of people are heavily reliant on one or more resources and when the resources' availability is unpredictable or constrained. People are likely to create a collective institutional system to address a resource problem if it occurs frequently, such as low or no catch, and if it affects a single community of users. Property-owning groups are social units with distinct membership and boundaries, shared interests, at least some interaction among members, shared cultural norms, and frequently their own endogenous authority systems. However, their nature, size, and internal structure vary widely. Examples could include neighbourhoods, small transhumant groupings, kin systems, extended families, sub-villages, tribal groups, or subgroups. Certain natural resources, like farmland, grazing grounds, and water sources, are traditionally owned by these groups. Different individual uses of a portion of the resources held under common property are not incompatible with corporate group property regimes.

For example, in many African countries with customary tenure systems, a group may possess a section of the farmland, and the group's leaders subsequently divide the land among other people or families. No one else has the right to utilise or profit from "their" plot as long as those people are cultivating it. However, keep in mind that the cultivator only has usufruct (use rights) and cannot alienate or transfer the land's ownership or use to another person. The land returns to the corporate ownership of the organisation after the present user stops using it effectively. The land-based property regimes in collective farms or agricultural cooperatives in the centrally planned socialist economies of Eastern Europe and the Soviet Union are essentially distinct from this type of common property. Members of the collective do not own the land in these entities as common property. Instead, the state owns the land. The significant reorganization currently occurring in Soviet agriculture, which is somewhat comparable to what has occurred in China, highlights the negative implications of state property, notably how it affects management practices for natural resources that are not considered common (or group) property. Restrictions on the output of labour applied to those resources are the most striking feature. The typical common property regimes in the developing world, in contrast to such state ownership regimes, are defined by group or corporate ownership, with management authority vested in the relevant organisation or its leaders. Some resources in the public domain (i.e., non-private land) are managed as common property in many developing nations, while others are handled as state property by the public sector and others are left open access without any management. Recognizing the distinctions between these regimes and designing programs appropriately is a good idea.

Given that there is a common misconception that the only motivation is to pillage and exploit natural resources, Common Property Regimes (CPR) incentives are crucial. In contrast, the accurate interpretation establishes a common property regime through collective ownership, where all group members' actions are governed by established norms and visible to all. It is not hyperbole to state that local group norm compliance serves as a powerful deterrent to antisocial behaviour in many cultures. Thus, a workable common property re□ime contains an inherent system of financial and non-financial incentives that promote adherence to established institutions and customs. Regretfully, in many contexts, those incentives and penalties have become ineffective, or dysfunctional, primarily due to internal processes that the groups were unable to manage or external pressures and forces. However, that doesn't change the fundamental idea that common property regimes have a cultural context that is consistent with and even necessary for their successful operation in a social context where individual adherence to group norms is the dominant ethic. A stable CPR structure requires the following principles:

- Clearly defined boundaries;
- Congruence between appropriation and provision rules and local conditions;
- Collective-choice arrangements allowing for the participation of most of the appropriators in the decision-making process;
- Effective monitoring by monitors who are part of or accountable to the appropriators;
- Graduated sanctions for appropriators who do not respect community rules;
- Conflict-resolution mechanisms which are cheap and easy of access;
- Minimal recognition of rights to organize (e.g., by the government); and
- In case of larger CPRs: Organization in the form of multiple layers of nested enterprises, with small, local CPRs at their bases.

Members of the group are prepared to forgo some of the advantages of using the resource individually in the face of resource availability uncertainty in exchange for the shared guarantee that the resource will be used in a more sustainable and equitable way. In order to accomplish the intended result, the □roup members are incentivized to execute specific behaviours by the institutional arrangements. However, the development of institutional arrangements necessitates a time commitment from the participants. Building institutions begins with coordination and information-gathering. Transaction costs will be incurred during the establishment and upkeep of institutions. These transaction costs are a component of the community decision-making process for common property regimes. Systems that manage communal resources for a lon□time can create institutional structures where the benefits of collective action are not outweighed by the transaction costs. Each member of the group depends on the other members of the group to reciprocate by adhering to the established management guidelines. In a circumstance involving collective action, an individual's decision will be based on how they balance the costs and advantages of different options as well as the probable results of each. Limited knowledge, the degree of opportunistic behaviour or self-interest users can anticipate from other resource users, and their discount rate all frequently influence an individual's behaviour. In certain circumstances, people could be motivated to use opportunistic tactics in order to go around the law and reap disproportionate advantages at the expense of others. This could involve corruption, free-riding, and rent-seeking. Establishing institutional mechanisms that can lower or minimise transaction costs and prevent opportunistic behaviour is essential to the common property regime. How to organised oneself is the main issue that group members of a common property system face. In other words, how to shift from a state of independent action to one of coordinated plans and □roup action in order to reduce joint harm and increase joint benefits. For the benefit of everybody, a sense of "commonality," dedication, and compliance must be developed. It is necessary to resolve issues with catch distribution and resource use assignment. Coordination issues arise in collective activity that are not present in other resource systems, like private property. Fishermen, for instance, must create regulations to specify how rights are to be exercised in order to coordinate their harvestin□ Rules offer incentive systems that influence interpersonal conflict or cooperation. Establishin□ practical protocols for keepin□ an eye on fishermen's behaviour, implementin□ punishments a□ainst nonconforming behaviour, and resolving disputes is crucial for the long-term maintenance of institutional arrangements. The physical characteristics of the resource, the rules, and the degree of adherence to the rules determine how simple and expensive the monitoring rules that were developed to structure the utilization activity are. To adapt to new possibilities, internal development, externalities, and institutional challenges, common property regimes and the institutional arran□ements that □o alon□ with them must be flexible. The process of creating an institution takes time and frequently relies on trial and error. For instance, a low crop may necessitate a change in allocation rules, or poor compliance may require a revision to a rule. Institutional structuring must be a continuous process to adapt to shifting circumstances.

In the majority world—a phrase I prefer over "developing countries" because the majority of people live there—threats to community-based property rights, including the commons, are severe, serious, and becoming worse every day, especially for local residents. The main issue is that most nation-states in the majority world disregard community-based property rights and assert control of large tracts of natural resources, including those inhabited by indigenous and other local populations. It is quite astounding how broad these assertions of state ownership are. The nation-state asserts control of considerably over half of the land base in many nations. For instance, the national government of Indonesia has legally designated an estimated 90% of the Outer Islands as state-owned forest land. Not all of these places are wooded. (Equating legal classifications with actual □round level situations should be avoided.) However, there may be approximately 70 million people living there, including more than 30 million indigenous people. Regardless of how long they have been occupying the land, all of these Indonesian nationals are legally considered squatters on public property. Many rural populations in Indonesia and much of the world are stewards and protectors of natural resources, such as carbon sinks and reservoirs of biodiversity, and they have valuable local expertise for sustainable resource management. Naturally, local cultures and conditions differ, and not all residents—including indigenous people—respect and preserve their natural surroundings. However, because they are all human, they need to have the fundamental right to take part in decisions that directly affect their lives and means of subsistence. Regretfully, rural populations in the majority world continue to be forcibly evicted from their ancestral homes, and international and national laws are routinely disregarded.

2.1.3. State Property Regimes

Under a state property regime, the state controls both ownership and usage. The materials may be available for use by both individuals and organisations, but only with the state's permission. State property regimes include military reservations, national (or state) parks, and national (or state) forests. It is possible to switch from state property to other kinds or the other way around. A common property regime at the village level was transformed into a state property regime, for example, when the government nationalized Nepal's village forests in 1957. Through government agencies, the state may either directly oversee the use of its natural resources or lease them to organisations or individuals, granting them usufruct rights over the resources for a predetermined amount of time. Examples include the "tree growing associations" that were experimentally established in West Bengal (and other parts of India) and comprised of groups of marginal or landless farmers who were allocated a plot of marginal public land to plant trees on. Although the group is granted usufruct rights on the land and ownership rights over its produce, the members are not awarded land titles.

2.1.4. Private Property Regimes

Private property is the most well-known type of property system. Despite the fact that most people consider private property to be individual property, corporate property is all private property even when it is managed by a group. Additionally, keep in mind the numerous responsibilities that come with private ownership of land and other resources; very few "owners" have complete autonomy over these resources. Right now, a few remarks on private property seem appropriate. Private property gives the lucky owner the socially and legally acceptable power to keep others out and to make them relocate. Those who believe that private property is the only source of ultimate wisdom, however, are responsible for a number of phenomena. First, the concentration of land ownership in the hands of a few wealthy families accounts for a large portion of the world's landlessness rather than a complete physical lack of land. Large swaths of Latin America are particularly affected. Second, private property is said to promote the "highest and best use of land." Since food crops are grown on poorer grounds and cattle ranching occupies large portions of

Latin America's greatest agricultural land, sceptics should be pardoned if they dispute that specific fact. As Proudhon stated, "theft" is not always what private property is, but many thefts have become private property, particularly in the western globe where European colonists took large tracts of land that were inhabited by indigenous peoples. In most contexts, the best land has already been privatized, while the worst has been left in the "public domain" as open access (res nullius), common property (res communis), or state property. It is not reasonable to expect common property regimes to manage resources that are highly variable and low in productivity, as well as to adapt and adjust to intense internal and external pressures when circumstances outside of that regime's boundaries make it impossible to do so. In other words, if classic adaptation mechanisms, like hiving off, are now barred by greater population growth beyond the boundaries of the common property system under consideration, the "internal pressure" of population expansion may be hard to address. Similarly, it is scarcely acceptable to criticises a people and their property system if private property and the fences that surround it hinder their traditional movements and those of their animals. Because private property regimes have the social and legal authority to keep out excess population and successfully fend off undesirable incursions through the state's power, they seem stable and adaptable. Common property regimes have reduced these abilities. Consider the consequences of primogeniture to understand the restrictive nature of private property. Since the eviction of younger sons—and even all daughters—is seen as a socially costless process, private property appears to be resilient and flexible; it "works." In such a situation, private property might "work" for the eldest son, but it might be more difficult to persuade those who have no claim to the estate.

Since common property is essentially "private" property for the group, deciding who should be excluded is a collective choice. However, people involved in the shared use of a resource are left with no choice but to deplete their capital when opportunities for the profitable and promisin□ exclusion of excess population have been eliminated by surrounding political, cultural, or economic events. But it's ludicrous to attribute this predicament on their inability to establish private property. Open access resources are not the same as common property. Similar to private person property situations, people have rights and responsibilities when it comes to shared (non-individual) property. The distinction between private and common property is more about how many rights and obligations are included or excluded than it is about the type of rights and obligations. The group's reluctance to remove unnecessary people when doing so would most likely condemn the displaced to famine is another area where they differ. In a way, rather than excludin□specific members from inheritance, the community decides to reduce its own quality of living. These areas ought to be "privatized" to preserve them, according to some. We question this oversimplified tenurial account of economic behaviour as there is no concrete proof that privatization lowers land exploitation while other economic incentives remain same. Furthermore, the attractiveness of privatization as a "solution" is dubious because it will just result in exclusion and population displacement to other locations (such as urban slums or other common property regions).

2.1.5. Private And Common Property Rights

The question of how private and common property relate to one another has sparked a lengthy discourse among legal and economic specialists over the definition, evolution, and relative merits of private and common property. The efficiency, equality, and sustainability of private property in comparison to common property are the topics of discussion. Both professions' scholarship has been defined by formulations that are taken up by successive generations of scholars without much effort to evaluate them empirically or look into their underlying assumptions. Each has a doctrinal component. Furthermore, the prevailing opinion in both fields has been that private property is obviously better than communal property. Many academics believe that modern instances of common property are relics from the past that will probably vanish in the 21st century (*see* Atran et al., 1993). However, the idea that private property is inherently better than common property has been questioned by recent studies.

2.1.6. Private Versus Common Property Legal Debate

Before the renowned English jurist Henry Sumner Maine published Ancient Law in 1861, Western jurists generally believed that the concept of property originated in ancient times when a single proprietor and his family occupied land (Grossi, 1981). Furthermore, the legal literature of the early 19th century upheld the superiority of individual property holdings to such an extent that the prospect of alternative property types prevailing on the European continent endangered legal theories regarding the genesis of social order. In addition to his own lengthy research in India, Maine drew on the groundbreaking work of William Blackstone (2016) and Georg Ludwig von Maurer's (1856) work on the Mark, a primitive Germanic village community. Maine came to the conclusion that "the forms of property that will afford us instruction will be those associated with the rights of families and of groups of kindred," and that "joint ownership, rather than separate ownership, is the really archaic institution." This sparked a rush of writings both disputing and defending his conclusion. Since significant political conflicts about the status of the numerous types of common property that still existed on the European continent persisted into the 19th century, the great argument was far more significant than just an academic one. The landed proprietor had a unique place in society because of a legal and political ideology that attributed the creation of property to the attempts of individual landowners to occupy it. These ideas supported the enactment of laws that permitted enclosures and the individual owners' occupation of communal assets, as well as the elimination of collective landholding rights. In contemporary legal studies, the definition of private property in relation to common property is still up for debate. For instance, the opening chapter of Ellickson, Rose, and Ackerman's (2022) most current property law textbook is titled "The Debate over Private Property." "The Problem of the Commons" is covered in the Chapter 2. Parts of Hardin's well-known 1968 paper on "The Tragedy of the Commons" are included in the latter chapter, but students are then asked the following questions: "Private property is often said to avert the tragedy of the commons." Does it, though? Who is responsible for enforcing property restrictions? Does the establishment and upkeep of a property regime present another type of "commons" issue? When Rose (1994) emphasises that a private-property regime as a system "has the same structure as a common property," she highlights the "kicker" in a clear division between private and common property.

2.1.7. Private Versus Common Property Economic Debate

Economists typically believe that common-property institutions have existed for a longer period of time than private-property institutions, and they attribute the development of contemporary Western societies in part to the shift from common to private property (North and Thomas, 1976). Due of the incentives associated with various types of property arrangements, the majority of economists believe that private property is a necessary component of economic progress. For instance, a farmer who owns his own labour, land, and other inputs is likely to observe a direct correlation between investments and the degree of long-term gain attained. On the other hand, a farmer who is a member of an agricultural production cooperative might only perceive a tenuous link between advantages and individual contributions. The more people in a society whose labour is only marginally related to their advantages, the more widespread the free-riding mentality can develop. Overall economic productivity will be low if everyone has a tendency to take advantage of other people's labour. However, a common-property system cannot just give rise to private-property rights. The presence and execution of a set of regulations that specify who is allowed to engage in what activities on their own initiative and how the profits from those activities will be distributed are necessary for the protection of private property rights (Ostrom, 2008). To put it another way, a property system cannot be established, monitored, or enforced without the help of rules and rulers. Even while certain regulations create incentives that significantly improve the well-being of the majority of participants in an economy, some people will always oppose changes because they stand to gain from the previous system or because they suggest improvements that will specifically benefit them. Making laws that favor some at the expense of others can also yield significant rewards for rulers. Therefore, it is predicted that both rulers and businesspeople will act in a rent-seeking manner. Therefore, many economists assume that common-property regimes are inefficient because of three issues. One is rent dissipation, which occurs when a resource's products are not owned until they are taken, and everyone competes in an ineffective race to take them before others do. The second is that if community owners were to create regulations to lessen the externalities of their mutual usage, they would likely have to pay substantial transaction and enforcement costs (Demsetz, 1967; Coase, 1960). The third is low productivity since there is little motivation for anyone to put in extra effort to boost their personal profits (North, 1990). Common-property regimes are likely maintained by leaders who are either backed by those who profit from these "archaic" regimes or who are unaware of the improvement in total economic welfare that would come from a shift to private property. "The only way to avoid the tragedy of the commons in natural resources and wildlife is to end the common-property system by creating a system of private property rights," Smith (1981) said, articulating a common policy prescription.

2.1.8. Confusions That Generate Misunderstanding

Three misunderstandings that impede scholarly communication have tainted the discussion of the relative benefits of private and communal property. Terms are given several meanings without elaborating on the connections between various elements. The misconception stems from the distinctions between:

- Common property and open-access regimes;
- Common-pool resources and common property regimes; and
- A resource system and the flow of resource units.

Theoretical and empirical advancement is slowed down by these three causes of confusion, which also make it harder to ascribe meaning to concepts.

2.1.8.1. Common Property and Open-Access Regimes

In a now-classic paper, Ciriacy-Wantrup and Bishop (1975) distinguished clearly between common property, where members of a clearly defined group have the legal authority to bar nonmembers from using a resource, and open access property regimes, where no one has the authority to bar anyone from using a resource (*see also* Bromley, 1998). Legal doctrine has traditionally regarded free-access regimes (res nullius), which include the well-known examples of the open seas and the atmosphere, as involving no restrictions on who is permitted to exploit a resource. If a resource is available to everyone, there is no reason for anyone to invest in upgrades or limit their use. One can anticipate that abuse and overconsumption will result from the absence of regulations governing permitted use if such a resource produces highly desired products. Although many fewer than assumed in the literature, some local woods, inshore fisheries, and grazing regions are indeed open-access resources. Effective laws defining property rights are by nature absent from certain open-access regimes (Dales, 2002). Either a nation-state does not contain the resources impacted by these open-access regimes, or no organisation has been able to establish rightful ownership. The result of deliberate public measures to ensure that every citizen has access to a resource under a political jurisdiction is another open-access regime. Although their official status is subject to the jus publicum principle, these resources are essentially freely accessible. Early in the 20th century, the Oregon and Washington state governments stepped in to stop local salmon fishers from creating regulations that would have set harvesting quotas and restricted admission (Higgs, 1979). In order to restrict entry and set harvesting quotas, fishing unions around the US coast attempted to organised inshore fisheries in the 1950s. The US Department of Justice prosecuted the fishing unions and found them in violation of the Sherman Antitrust Act, despite the fact that their efforts could not have had a significant effect on prices because of the existence of an active worldwide market for fish (Johnson and Libecap, 1982). Because of the government's efforts to stop local fishing organisations from establishing common-property regimes inside their political jurisdictions, US inshore fisheries have essentially been open-access resources throughout the most of the 20th century. However, both the federal government and state governments have recently changed their positions and aggressively looked for ways to establish co-management in inshore fisheries (*see* Wilson, Yan, and Wilson, 2007). The ineffectual exclusion of nonowners by the body granted formal ownership rights leads to a third kind of open-access regime. Ironically, the amount and scope of locally available resources that are essentially open access have increased in many developing nations as a result of the previous misunderstanding between open-access and common-property regimes. Although they had developed over many years in various parts of the world, common-property systems governing access and harvesting from nearby streams, woods, grazing areas, and inshore fisheries were rarely granted official status in the legal codes of recently established nations (Wiersma, 2005). Many developing nations nationalized all land and water resources that had not yet been registered as private property as concern over the preservation of natural resources grew in the 1960s. Although the national governments lacked the financial means and manpower to adequately oversee the use of these resources, the institutional arrangements that local users had developed to restrict access and use lost their legal substance. As a result, resources that had previously been subject to a de facto common-property system enforced by local users were reverted to a de facto open-access regime while being transformed to a de jure government-property regime. State control has typically proven to be less effective and efficient than control by those directly affected, if not disastrous in its consequences, when resources that were previously controlled by local participants have been nationalized (Curtis, 1991; Hilton, 1992; Panayotou and Ashton, 1992; Ascher, 1995). In Thailand, Niger, Nepal, and India, the negative consequences of nationalizing forests that were formerly managed by local user groups have been extensively documented. Comparable outcomes have been observed for inshore fisheries that have been transferred from local management by inshore fishermen to state or federal entities.

2.1.8.2. Common-Pool Resources and Common Property Regimes

A second terminological issue makes it especially challenging to resolve the issues that arise from conflating open-access and common-property regimes. A sort of economic good that is more appropriately referred to as a "common pool resource" is commonly referred to as a "common-property resource." Two characteristics that are crucial for economic activity are shared by all common-pool resources:

- It is expensive to prevent people from using the product by means of legal or physical restrictions; and
- One person's advantages deduct from those of other people (V. Ostrom and E. Ostrom, 1977b; E. Ostrom, Gardner, and Walker, 1994).

By identifying a class of goods that match these two characteristics, researchers can pinpoint the fundamental theoretical issues that arise whenever multiple people or groups use these resources for a prolonged length of time. The idea that goods with similar characteristics tend to have the same property regime everywhere is strengthened when the term "property" is used to describe a particular kind of commodity. Similar to public goods, common-pool resources face the challenge of creating institutional or physical barriers to excluding recipients. The tremendous desire to profit on the work of others will result in a less-than-ideal investment in resource improvement, usage monitoring, and the punishment of rule-breaking behaviour unless strategies are developed to prevent unauthorized users from doing so. Second, the units of resources or products from the common pool have the same characteristic as private goods: the quantity available to others is reduced by the consumption of one individual. Therefore, unless harvesting or use limitations are established and implemented, common-pool resources are vulnerable to issues of congestion, misuse, and possible destruction. Apart from these two characteristics, specific common-pool resources vary in numerous other aspects that impact their economic utility, such as their dimensions, form, and efficiency, as well as the worth, timing, and consistency of the resource units generated. National, regional, or local governments, community organisations, private citizens, or businesses may own or use common-pool resources, or anybody with access may use them as open-access resources. Although each of the major categories of property re□imes has unique benefits and drawbacks, they occasionally may be based on comparable operational guidelines for resource access and usage. There are instances of both successful and unsuccessful attempts by governments, community or□anisations, cooperatives, nonprofit or□anisations, and private persons or businesses to control and manage resources from a common pool (Bromley et al., 1992). Therefore, as will be explained below, common-pool resources are not automatically associated with common-property regimes or any other specific kind of property rule. Furthermore, as common property agreements are essentially share contracts, they are susceptible to moral hazard issues and possible opportunistic behaviour.

2.1.8.3. Resource System and The Flow Of Resource Units

Resource systems and a flow of resource units or benefits from these systems make up common-pool resources. Over time, a flow of resource units or benefits is produced by the resource system, also known as the stock or the facility. Lakes, rivers, irrigation systems, groundwater basins, forests, fisheries, and grazing regions are a few examples of common-pool resource systems. Facilities built for shared use, like mainframe computers and the Internet, can also be considered common-pool resources. Common-pool resources include things like water, lumber, fish, feed, medicinal herbs, central processing units, and connection time. Rules that restrict access to the resource system and other regulations that restrict the quantity, timing, and technology of removing various resource units from the resource system are necessary to design property regimes that successfully permit sustainable use of a common-pool resource.

2.1.9. Superiority of Different Regimes

Private property rights are typically regarded as the more effective institution from an economic perspective. This is because of the externalities that arise in relation to common property and open access, as well as the incentives that they generate. Take, for instance, a farmer who is starting to cultivate his property. If the property is privately owned, there is a direct correlation between the money made and the investments or labour inputs used. The relationship between individual contributions and returns will be weaker under other property rights regimes, such as common property or open access. When there is a chance that someone else will take advantage of the farmer's labour, it makes sense for him to put in less effort as well. On the consumer side, the opposite is true. Since the benefit of restraint will mostly benefit other hunters in the same area, no hunter has any motivation to limit his own hunting if private rights are missing in a particular resource, such as game (cf. Demsetz, 1967). You might say first come, first served. If there are no exclusive rights in fish stocks, the position with regard to fisheries is similar. Depletion and eventual extinction may result. An article titled "The Tragedy of the Commons" has had a significant impact on public perceptions of both common property and open access. Hardin (1968), the author, used the example of grazing area that was accessible to everyone. Because of this, we may anticipate that every herdsman will make an effort to maintain the greatest number of animals there. This system would ultimately destroy the pasture because the cost of each extra animal is split among all the herdsmen. The remedy that is frequently promoted is the creation of exclusive, individual rights. The shift to a private property rights framework will internalize the externalities. For instance, if the land is divided into property units, the cost of resource utilization and the profits of management methods will devolve on the individual property owner. Numerous studies of economic history have examined the processes of development in various societies. An institutional shift from an initial state of free access or common property to regimes of private property rights is a commonality throughout many of these studies. Examples of such topics have been the creation of mining rights in California, the establishment of private land ownership during the colonization of the United States, the English enclosures, and the beginnings of hunting rights among Indian communities in modern-day Canada (Demsetz, 1967). Similar research has also been done on Swedish conditions. Among other things, Lundgren (1987) examines how the Lule River valley's hunting and fishing rights have been institutionalized. As a result of the trade that evolved between the Sami and the farming people during the mediaeval era, exclusive rights to these resources were established. The land redistribution reforms, which signaled a slow shift from common rights regimes to individual property units during the 18th and 20th centuries, are another example of changes in the Swedish property system. In conclusion, free access and common property are perceived as encouraging excessive consumption since consumers disregard the expenses associated with their own use that are faced by other users. However, because the person implementing the management measures is unable to fully profit from them, the results of those measures are insufficient. If all users could agree, for example, on the use limitation, this "tragedy" may be prevented. However, the agreement could not be used against new users even if this were feasible. Signing contracts with all current and prospective consumers would be prohibitively expensive.8.

The resource will be more valuable to those who are not constrained by any agreements if the current group is successful in agreeing to limit its own use. The replacement of open access and common property regimes with private rights appears to be the answer in both theoretical models and empirical research.

2.1.10. Reasons Why Not All Resources Are Under Private Property Rights

I mentioned in my introduction that Swedish property law contains rules that are based on open access or common property, even though it seems that individual rights are preferable to open access and common property regimes. Examples of resources that lack exclusive and individual rights include the atmosphere and marine resources like fish and minerals. It is obvious that the advantages of individual rights must be weighed against the different costs associated with individualization. Two such factors—the existence of institutional costs and what are known as economies of scale—will be discussed in this section.

2.1.10.1. Institutional Costs

There are costs associated with establishing and preserving exclusive land rights. Reference points, registers, maps, and the like must be used to document the territorial subdivision. It is necessary to mark and keep an eye on boundaries. Violations must be preventable and subject to legal action through public sanctions.

All of these costs are associated with the institution. A resource-demanding rights regime must have positive land use impacts, such as increased land values, to balance out its institutional costs and be economically justified. The benefits result from altered incentive structures and reduced transaction costs for agreements pertaining to transfers, rights grants, etc. The graphic below, which depicts how institutional costs and land use effects evolve on the margin, serves as an illustration of this point. The amount of resources required to establish and preserve property rights in land is shown by a shift to the right on the horizontal axis (Figure 2.2).

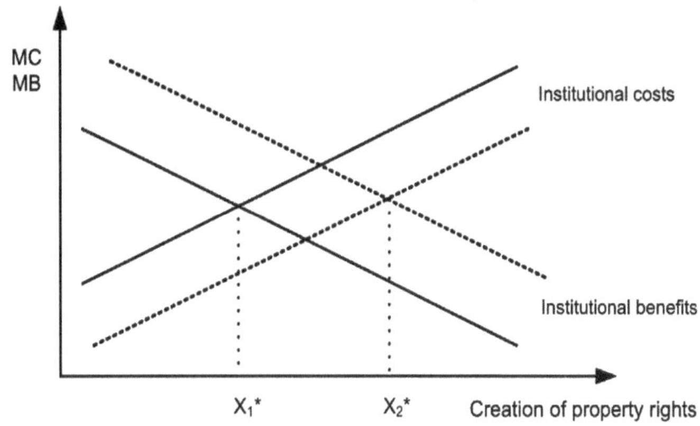

Figure 2.2. *Marginal benefits and costs in relation to the creation of property rights.*

Source: Ekbäck (2000).

It is assumed that the advantages of the new property rights framework will grow, albeit more slowly. This is because each input factor's marginal productivity is decreasing. As a result of rising opportunity costs for resources diverted from other economic production, institutional costs instead rise somewhat. Therefore, an effective institution should be built so that the marginal difference between the higher land use value and the institutional costs is equal. This happens in X1* in a hypothetical scenario. The mar□inal utility function may migrate outward in this scenario as a result of increased demand for land resources brought on by population growth, altered production methods, or the emergence of new markets. This change suggests that land resources are valued and supports the employment of more resources to establish property rights. A new, efficient equilibrium is reached in X2* if new technolo□y lowers the costs of the new rights regime. This will also cause the marginal cost curve to shift outward. Complete privatization of rights in all resources can never be economically efficient due to institutional expenses, as Fi□ure 2.1 illustrates. Rather, the equilibrium point indicates that some resources will either be owned jointly or not at all be subject to exclusive rights. Depending on the technological and physical characteristics of the resources in question, the institutional costs shown in Fi□ure 2.1 can fluctuate in size. In relation to the development and upkeep of private rights, common pool resources—such as fish stocks, □roundwater aquifers, or □ame animals that wander vast swaths of land—are characterized by extremely high institutional costs. Exclusion costs, or the expense of preventing others from exploiting the resource, make up the majority of the high costs (Ostrom, 1999). Common property rights or even no rights at all (open access) may be more effective than individual ri□hts in situations like this due to si□nificant exclusion costs. When facilities produce positive externalities in their surroundin□s when it is costly or difficult to exclude other people from these advantages, a unique scenario occurs. One type of facility is a land drainage enterprise. Therefore, there is a chance that someone else will take advantage of the person running the operation. The inability to finance such facilities could result in an unreasonably low level of output. Common property rights might be advantageous in such circumstances.

2.1.10.2. Economies of Scale

There are increasing returns to scale in the technology of some land uses. In other words, the operation necessitates significant upfront investments, meaning that the fixed costs of production are substantial. At least until the facility's capacity limit is reached, more benefits can be generated at a comparatively low cost once these investments have been made. One business that uses economies of scale is aquifers (Ekbäck, 2007). Land surveys, test drilling, purification plants, etc., might have very large fixed expenses. Drilling a well deep into the bedrock may reveal areas of fissure in the aquifer, which would let the facility to feed many properties. Regarding irrigation facilities, the situation is comparable (Ostrom, 1999). Other facilities that clearly demonstrate economies of scale are roads, sewage systems, and landing stages. In most cases, multiple people must work together to take advantage of the economies of scale associated with such operations. In the event that the rights in question are private, discussions and agreements pertaining to facility building and cost allocation must be reached between the parties. High transaction costs may result from this. There will be no need for each participant to negotiate a separate agreement if the rights are instead shared property. Agreements that demand resources are not necessary to take advantage of economies of scale. However, the costs associated with group coordination and decision-making will also be related.

2.1.11. Transitions Between Property Regimes

Why do societies redistribute property rights? The enclosure of European common fields, the partition of hunting lands by Montagnais Indians, the initial distribution of open-access property that most likely occurred at the genesis of property rights, and other transitions between property regimes are fundamental to the economic study of property rights. Why do these changes take place? Harold Demsetz's "Towards a Theory of Property Rights" provides the standard explanation, which holds that societies choose new property regimes when an external shock changes the costs and benefits of an existing regime to the point where it is less efficient than the new one. Demsetz gives the example of the Montagnais. Externalities related to open-access hunting arose when the commercial fur trade was established as a result of colonial settlement. Demsetz contends that as a result, the Montagnais started delineating their hunting areas and granting exclusive hunting privileges to tribe members in certain locations. Because it suggests that property rights will eventually be redistributed in the direction of efficiency, the Demsetz story is a pleasant one. As others have pointed out, the problem with this explanation is that it doesn't explain the precise mechanism via which the shift takes place. It is not enough to presume that, like other things, property rights are produced in a way that adapts to the shifting costs and benefits of production. The development of a property right is invariably a group effort, regardless of whether it is done formally by a government or informally by individuals working together to create a social norm. The shift from less efficient property regime A to more efficient regime B cannot be attributed to a single entrepreneur. Only if there is a way to convert the collective cost-benefit balance into a sizable sum of the individual cost-benefit balances will a society be able to advance from point A to point B. It is unclear whether such a mechanism exists because switching between property regimes is probably going to be expensive. It will be necessary to assign rights under the new regime and value those under the previous one. The collective action issue that arises throughout the transition project must be resolved by the participants. Without a society's means of overcoming these barriers, no shift could proceed, yet the Demsetz account provides none.

Given this challenge, a different perspective on changes in property regimes emerges, according to which societies redistribute property rights whenever an external political realignment gives a dominant group a bigger piece of the pie. Given our limited knowledge of the Montagnais, one could also tell this type of story about them. It's possible that colonial settlement changed the balance of power within the tribe. For example, it gave tribe members who interacted with the settlers more authority, allowing one group to snuff out the others by splitting up the hunting grounds. Due to their lack of writing, the Montagnais left no traces that would indicate which story is correct. Since there are no efficiency implications in either direction, this is a darker scenario than Demsetz's. It implies that there should be no discernible pattern to the long-term shifts between property regimes, which should happen in both directions—inefficiently and efficiently. One advanta□e of this approach is that it outlines the process via which shifts take place. However, compared to the Demsetz account, it seems to be far less consistent with experience. Transitions between property re□imes do appear to have tended towards efficiency over the long term. In early modern Europe, for instance, enclosure was unidirectional. There was no tendency towards the dis-enclosure of property owned wholly by a single individual, nor even a patchwork of enclosures and disenclosures, despite the fact that there was a lot of opposition to enclosure and even several statutes that prohibited actions related to enclosure.6. Generally speaking, it is simple to recall instances of resources where exclusive rights were initially established when their value rose, as the Demsetz story foretells, but it is difficult to recall instances of the reverse—that is, when a resource's value rose in tandem with the renunciation of exclusive rights or when exclusive rights were established after its value fell. We would anticipate more inefficient transitions between property re□imes if they took place for non-efficient reasons. Thus, the Demsetz tale is more in line with the actual data than its rival, but it is lacking two important details. It must first explain how a society mi□ht □et beyond the barriers that could prevent it from moving towards a more effective property law. How is the issue of collective action resolved by transition participants? How do they manage the expenses associated with assigning new rights and valuing existing ones? Second, it requires an explanation of the process by which political action is converted from efficiency. Although a beginning state and an end state are described in the Demsetz tale, no means of connecting them is provided. Why do chan□es that □o towards efficiency seem to be able to get enough political support to be implemented, but changes that move in the opposite way do not?

2.1.11.1. Obstacles To Transitions

Assume that there are two property systems: A, which is less efficient, and B, which is more efficient. Assume that a society has A at the moment and that its members are aware of B. Why wouldn't society just go from point A to point B?

2.1.11.1.1. Collective Action

The collective action issue raised by the change from option A to option B has been the focus of Demsetz's most persistent criticism. Although it takes time and effort to establish and implement any new property rights system, the system itself seems to be a public good that will benefit everyone in a given location, including those who declined to assist with the organizational work. James Krier has referred to the outcome as "the tragedy of the commons, part two." Everyone is motivated to take advantage of other people's organizational efforts. The original shift between property systems, which occurred with the earliest human attempts to establish property rights, is where the issue is most pronounced. What was the origin of property? However, in a society where a property system is already in place, the collective action problem also exists with regard to property rights in any particular resource. Establishing property rights in the fisheries would benefit all fishermen, but someone has to invest the resources required to make it happen. But who would, knowing that he would reap the rewards of privatization, spend his time fishing and waiting for someone else to do it? A few arguments suggest that this collective action issue might not be as serious as it first seems, but none of them completely eliminates the challenge. First, because members of close-knit organisations trust one other enough to act without worrying about others leaving, they often develop mutually beneficial rules. These societies might have been the ancient tribes who developed the first property systems. Certain co-users of a shared resource, such as a fishery, might be grouped together in small enough groups to have mutual trust. However, other changes—such as enclosure—have occurred on a very broad scale and frequently necessitated the cooperation of strangers, indicating that there is still a class of transitions that cannot be described by the existence of small-group norms. Second, property conventions can (at least in theory) develop naturally even in the absence of trust. This is demonstrated by the equilibrium of a repeating game in which pairs of resource claimants decide between tactics of taking the resource and allowing the opponent to take it. This could be the most plausible account for how property rights came to be. However, it appears to have minimal effect on the change from an old property rights system to a new one, albeit the latter class of transitions might also be explained by a game that is constructed differently.

Third, compared to the initial construction of a property system, the collective action problem may not be as si□nificant with re□ard to individual resources within an established property system. For instance, fishermen may be able to work together to externalize a large portion of the cost by convincing the □overnment to create and enforce property ri□hts in the fishery at the expense of taxpayers, even if they are unable to collaborate on the costly project of creating and enforcing their own system of property rights. For each given resource, the cost of transitions is eliminated by an existing property system. The collective action problem does not go away, though, because that expense is still incurred for a gain that a group will share. Furthermore, the early construction of a system of property ri□hts still faces si□nificant challen□es related to collective action. Lastly, it is not necessary for a system of property ri□hts to be a public benefit in the relevant sense, despite popular belief. By simply refusing to enforce those people's claims, the organizers of a property system can deny those people the benefits of the system. We just don't □ive a fisherman a share if we are splittin□ up a fishery for the first time and he refuses to assist. Since insiders cannot stop outsiders from supportin□ a property system, it is typically a public benefit. However, in the first division, it is conceivable to penalize individuals who attempt to avoid the organizational effort. When the British colonial authority of New Zealand restructured Maori land tenure in the latter part of the 19th century, such fines were really applied to nonparticipants. However, this merely moves the issue back in time and does not solve the collective action dilemma. The same collective action problem applies to the decision to deny nonparticipants legal rights under a new property system, which in and of itself necessitates some coordination. Before the community can legitimately threaten to refuse a share to nonparticipants, a specific percenta□e of participation is required. What is the process of that coordination? Therefore, collective action continues to be a barrier that could prevent changes in property regimes. That's what some closely observing transitions have thought. One of the judges of the Native Land Court in 19th century New Zealand, Frederick Maning, was convinced that the Maori would much rather own land in fee simple and that the reason they had not abandoned their traditional usufructuary system prior to the arrival of the British was because land ownership was so fragmented that it was impossible to take collective action. A similar tale of divided property rights and the associated organizing costs that are impeding Russia's shift to private property has been recounted more recently by Michael Heller. At least occasionally, collective action seems to be a major barrier.

2.1.11.1.2. Transition Costs

The administrative expenses associated with determining the worth of each person's rights under the previous system and identifying equivalent rights under the new one may be a second type of barrier to a transition between property regimes. These expenses differ from the well-known costs of setting up and implementing a property system, such as those associated with surveys, title registration, law enforcement, courts, and the like. Changes in property regimes come with extra costs, which are not typical. For instance, dividing up a commons among a tribe of 50 individuals seems straightforward in theory. All you need to do is give each person 2% of the land. However, it is unlikely that their resources and population will be uniform. What if the 50 tribe members use varying amounts of the varied resources found in their commons because they have varying endowments in terms of size, strength, diligence, and other factors? How do we value each person's portion? Do I deserve less than 2% as a skinny but gluttonous member of the community? Over 2%? And how can we convert that into real resources after we agree on the value of my share, say precisely 2%? Will I receive 2% of the surface area? What if some of the parcels are nearer the river or have more trees? Should those who receive a bunch of rocks receive a side payment from the folks who live close to the water? What are the side payments? Accurately answering all of these issues would take a significant amount of time and effort, which might easily outweigh the benefits the tribe would receive from partitioning the commons. When switching from one existing property system to another, the costs of valuation and allocation are probably going to be significantly higher. A certain amount of path dependence results from an initial distribution of property rights. Reallocation could be too costly to overcome once rights have been assigned. Take, for example, the historical shift from a usufructuary to a fee-simple structure of land ownership, which I shall go into more detail about shortly. Let's say I am allowed to harvest berries from a specific tree. How much is that worth? We cannot look at the market price since there isn't a well-developed market for my right to gather berries—at least not among many indigenous tribes. The value of the berries, the tree's yield, its position, and other factors all affect how much my property right is worth. After we have all that information, we can figure out how much I will lose if the person who gets the area where the tree grows is given the opportunity to harvest berries instead of me. We can now proceed to the next tree, and the one after that, and the one after that. Furthermore, determining the worth of my property rights is just half the task. Let's say we determine that my ri□hts are worth X in total. I have the ri□ht to an area of land valued at X. What size will it be? Where will it be, too? Will there be enough property near the river for others whose trees produced more berries than mine if I purchase a parcel close to the river? If not, should I be mandated to pay someone on the side? What is the amount? And where can I find the funds? Transitions between property regimes that would normally increase community wealth are hampered by the set of one-time administrative costs associated with valuing and allocating rights, similar to the collective action problem. Clearly, societies do get past these challenges occasionally, but how? There have probably been instances where civilizations have not been able to get over these barriers, and transitions have not taken place when the Demsetz account would have predicted that they would. For instance, traditional property regimes are still in place in a large portion of the Third World today, even though outsiders with economic training think they are significantly less effective than alternative possible regimes. What sets civilizations capable of handling shifts to more effective property systems apart from those that cannot?

2.1.11.2. Overcoming the Obstacles

We might start by looking at a collection of property regime transitions that have occurred over the past few centuries in most of the world to see how societies have managed to get over the challenges of collective action and administrative expense. Nowadays, land is usually separated spatially, and with rare exceptions, the owner of a piece of land is thought to control all the resources inside the parcel's boundaries. However, it wasn't always the case. European colonists discovered property systems in the majority of the non-European world, where resources were distributed functionally rather than spatially. Instead of owning a specific area, a person or family would have the right to use specific resources that are dispersed among several locations. For instance, among the Maori, one can be the sole owner of the right to farm a certain piece of land, fish in a particular area of the ocean, or catch birds in a particular tree. When a person was granted this kind of privilege, it did not follow that they would also be granted additional rights in the same area. For example, one family may utilise the same tree for berry picking, another for fowling, and so on. Furthermore, using one resource in one location did not negate the ability to use other resources in other locations. A family may be entitled to one area for eel catching, another for cultivation, and a third for resting. Like the majority of indigenous peoples over whom Europeans imposed their rule, the Maori had a sophisticated property system, but it was one in which property rights were not confined to a specific geographic area. A large portion of Europe had comparable nonspatial property arrangements prior to enclosure. Participants in common fields had the right to use certain resources in various locations rather than having complete control over all the resources within the designated bounds. During the farming season, a person may have the right to cultivate multiple strips of land spread across multiple fields, the right to graze a specific number of animals of a particular kind in specific pastures at specific times of the year, the right to collect nuts in one location and twigs in another, and so forth.17 Although a sizable portion of the productive resources were distributed by use rather than location, European property rights were not solely distributed based on functionality. This is because, as is the case now, a large portion of the land was separated physically. Much of this type of nonspatial allocation was eliminated between 1500 and 1900 when Europe and the colonies to which the majority of Europeans emigrated underwent a significant reorganization of property rights. Take, for instance, the British Empire and its successor. In Britain, the modern individual farm landscape was established by enclosing communal fields. The indigenous property systems that had previously distributed the resources within those blocks and divided the blocks spatially for settlement were eradicated by the British and American governments as they acquired or simply occupied large blocks of land in North America, Australia, and New Zealand (they did both). In order to restructure property rights in the numerous places that remained where indigenous people still distributed resources based on use rather than space, Britain and the United States established programs in the second half of the 19th century that were modelled after the previous enclosure commissions. In the 1860s, the colonial government established the Native Land Court in New Zealand, which was the first of these. Britain quickly established a comparable organisation to redistribute property rights in Fiji. After years of discussion over whether the best way to integrate the Indians was through the spatial distribution of their land, the Dawes Act of 1887 authorized the same type of reorganization across the western United States.

The late 19th century reallotment plans in the British and American empires, as well as the local specifics of enclosure in Britain, differed, but the programs had several things in common. Officials from the government were supposed to gather evidence, determine each claimant's use rights, and then allocate each claimant a zone of land worth the total of those rights. By increasing the land's productivity and ease of sale, the schemes all aimed to raise total wealth. And that's probably what they all did. After enclosure, agricultural rents typically increased in Britain, indicating that tenant farmers were expanding. Allotted Indian land was (and remains) more productive than unallotted land in the United States. Due to allotment, a significant amount of land was transferred from Indian to white owners, who were previously unable to buy it and likely had a higher value for the land (or at least could afford it). These schemes are classic Demsetz tales: individuals changed from a functional to a spatial allocation of property rights when new agricultural methods and population increase made it more economical to do so. Since this is a list of transitions that actually happened, it's easy to miss two additional characteristics that they all have. All would have been extremely costly—possibly too costly to have happened—had they been meticulously carried out. If everyone's interests had been carefully valued and allocated, the transition from a functional to a spatial system of property rights would have been an administrative nightmare, and the expenses could have easily offset any efficiency improvements. Additionally, all of these plans ought to have been vulnerable to serious issues with collective action. The cost of participation was high for the property owners, who had to spend time testifying and occasionally pay for surveys, attorneys, and court costs. Why didn't property owners avoid the paperwork? How could a sufficient number of property owners band together to legitimately threaten to bar anyone who didn't want to take part? Another pair of shared traits has the answer. Each of these initiatives had important distributional repercussions. The shift from functional to spatial property rights had certain winners and losers. The anticipated benefits for the winners were sufficiently substantial to encourage them to shoulder an excessive amount of the reorganization's administrative expenses. Furthermore, the distributional effects were not arbitrary. All of these plots took place in governmental systems that were hierarchical and mainly split into two classes. In general, the colonies had both settlers and native people, while Britain had large landowners and everyone else. In every transition, the wealthy and influential emerged victorious. As a result, an oligarchy was frequently pitted against a greater number of comparatively helpless farmers in the political economics of the transition. The people in charge of the governments that made the decisions about whether or not reorganization would occur were also the major beneficiaries of it. These initiatives made it easier for property rights to be redistributed by favoring the wealthy and powerful.

The efficiency advantages from these changes would not have had to be dispersed so unequally in a world without administrative overhead. The winners could have been anyone. However, the managers of the transitions had to take certain short cuts due to the enormous administrative expenditures. In order to make the change possible, they needed to implement some general guidelines that would keep the costs of assignment and valuation minimal. However, which □eneral □uidelines? Different types of people would have benefited from different regulations. Because the individuals in charge of the switch had the authority to select the rules of thumb, it should come as no surprise that they selected those that would benefit themselves and their allies while harmin□only defenseless stran□ers. This is where political hierarchy became si□nificant. For instance, the use rights of the poorest commoners in Britain were frequently worthless. This resulted in si□nificant administrative cost savin□s. It may have been more expensive to price the right than the right itself was worth when the property right in question was as small as picking up grain scraps after the harvest. If every cent had been accurately accounted for, the cost of enclosure might have been too high to justify the effort. Naturally, though, that did nothing to comfort the impacted commoners, who viewed enclosure as the seizure of their main source of income. Regardless of the actual number of people with resource rights within the block, Native Land Court justices in New Zealand developed the practice of registering land blocks to a maximum of 10 Maori. By lowering the number of Maori with the ability to protest, this type of corner cutting undoubtedly served the dual goals of lowering the expense of identifying the owners and enabling land sales to British settlers. Once more, though, the thousands of Maori whose only substantial asset abruptly disappeared were not particularly concerned about the decrease in transaction costs. Therefore, in spite of the overall benefits of these chan□es, they were all fiercely opposed at the time, in such depressing terms that even the most strict utilitarians took a moment to consider the losers. The distributional impacts of enclosure and the concurrent colonial programs were similar to those of free trade today, with concentrated losses for some and dispersed gains for the majority. However, because the poor were poorer, concentrated losses for them had a far sharper edge than they do now, makin□the conflicts more severe than the current fi□ht over free trade. In their day, these shifts from a functional to a geographical system of property rights were among the most painful and polarizing political issues in Britain and its colonies and former colonies. These conflicts are lon□since over and are now only contested by historians. Thus, a strong oligarchy can get around the administrative expense issue by adopting a less stringent approach when valuing and allocating the majority's property rights. As a society gets more equal, the majority has the ability to obstruct the change, making this solution less achievable. If a wealth-enhancing shift between property systems is to take place at all, it occasionally necessitates some roughness in the assignment and valuation of rights. This roughness will result in an unequal distribution of the transition's profits and costs in a hierarchical political framework, favorin□ those in charge of the transformation.

An oli□archy also makes a si□nificant stride towards solvin□the collective action problem when it skews the payoffs in this way. A smaller group of people who expect disproportionately big gains from a transition will be more motivated to work together to organised it than would a larger group of people who expect more evenly distributed rewards. Everyone's participation is not necessary for the oligarchs. To effectively threaten to deny nonparticipants property ri□hts under the new system, they simply need a sufficient number of participants. The fewer people needed to create that credible threat, the more concentrated political power will be and the more unequally the transition's benefits would be delivered. Therefore, this is empirical proof of a mechanism that allows a change from one property regime to another in order to get around the challenges of administrative expense and collective action. It is the process that enabled some of the most si□nificant chan□es in contemporary times, albeit it may not be the only one that might have occurred. The general (and testable) hypothesis that fewer equal societies are more prone to experience property regime shifts is supported by this analysis. Even though widespread European settlement and trade appear to have given them every incentive to do so, many indigenous societies with political structures more egalitarian than the pattern of settler-native relations that replaced them chose not to reorganize their traditional property systems. This provides additional evidence in favor of this claim. Invoking mystical ideas a sacred connection between indigenous people and the land, a sense of belonging to the land rather than owning it, etc.—along with the claim that Europeans viewed land as a commodity to be bought and sold is the standard explanation for the persistence of traditional, nonspatial property systems. On both sides, the contrast is undoubtedly overdrawn. Despite their affection for the land, many indigenous people in numerous locations were very willing to sell it in order to get all the practical manufactured products that would not have been possible otherwise. In contrast, romantic ties to their homeland were not uncommon among Europeans and their descendants. The necessary component for colonial and postcolonial transitions between property systems may have been the unequal distribution of political power between colonizers and colonized. The costs of transition might have been too great in the absence of that imbalance.

Chapter 3

Property Rights and Development

Contents

OVERVIEW

Individuals are the first to develop. Think about a tiny business owner. An entrepreneur must think of ways to boost production, buy or create new technology, invest in staff and equipment, and make long-term plans in order to expand their company. Among other factors, the resources used, the availability of markets, and societal stability will all affect the entrepreneur's eventual success. However, the ability to create, make choices, and pursue objectives rests with the person. The most likely development policy to achieve sustainability is one that unlocks this power and encourages positive individual behaviour. What can inspire someone to have an entrepreneurial spirit? A strong property rights framework can do this more than any other single component. Property owners are more willing to invest effort and take risks. As former Treasury Secretary Lawrence H. Summers noted, "No one has ever washed a rented car in the history of the world." Real estate ownership represents opportunity, accountability, and financial independence, enriching society by fostering innovation, encouraging investment, enabling trade, and improving governance. This essay examines how property has transformed the political and economic landscape of industrialized nations and its potential benefits for developing countries. By shaping incentives and strengthening societal institutions, property lays the groundwork for development, liberty, and well-being. Improved governance and regulatory reform lead to better property systems.

3.1. PROPERTY AND FATES OF NATIONS

The contrasting histories of England and Spain between 1500 and 1700 highlight how crucial property rights are to long-term growth. Profits from trade in the Low Countries and New World wealth allowed the Spanish kings to establish and sustain an empire. They were unable to establish a domestic climate that would support future expansion, though. In Spain, the monarchy failed to establish a system of agricultural property rights and instead imposed taxes mostly on sheep, which roamed the countryside. Under pressure from the cost of war and the Low Countries' uprising, the Crown unilaterally repudiated its debts and seized more and more property from economic, industrial, and agricultural interests. Long-term deterioration resulted from these constraints discouraging innovation and constructive activities. Like the Spanish monarchy, the English monarchy granted monopolistic privileges to royalty, supervised guilds that controlled production, and showed little regard for private property. Although they never gained the complete authority to do so, the Tudor and Stuart kings also attempted to increase their tax income. On behalf of the merchants and gentry, Parliament battled the Crown for the ability to set taxes, the abolition of monopolies, and a legally based system of property rights. The government would continue to receive cash as a result of Parliament's eventual win, but it would now protect property rights rather than take property away. In addition, economic and civil liberties were extended. The Industrial Revolution was made possible by the 18th century growth of competitive markets and the bolstering of financial incentives for invention. Many aspects of the political, economic, and institutional frameworks of the European powers were effectively transplanted to a number of their colonies. Regardless of European cultural influences, nations that lacked strong property rights and other essential market economy institutions saw slow or negative economic growth, while those that did enjoyed prosperity. The Korean experience demonstrates that this is not solely a Western phenomena. At the start of the Korean War, South Korea and the North had comparable endowments, but their economic systems were drastically different. As a result, the South prospered by upholding property rights and promoting markets, whereas the North, which outlawed private property, suffered. Since property is ubiquitous and usually secure, its function as a basis for economic progress is sometimes taken for granted in the developed world. In contrast, many developing nations are severely impacted by the lack of property. A disproportionately high amount of informal economic activity is usually how these consequences show up. People still own things in the unorganized sector, but they are not legally protected as property because they are not officially recognized. "They own homes but no titles, crops but no deeds, and companies but no incorporation statutes." As a result, those who work in this industry—primarily the impoverished—cannot use their possessions to prosper. The expenses of informality and the quantity of undercapitalized assets held by the poor have been revealed by research conducted by the Institute for Liberty and Democracy. A home with a title is worth nine times as much in Lima, Peru, as one without one. In Lima, a street seller who wants to obtain an ambiguous, illegal right to occupy their place of business must spend between 98% and 495% more for real estate than formal merchants (Soto, 1989). Eviction is still a possibility for informal traders and homeowners, who make up the bulk of Peru's housing and commercial establishments. Despite these expenses and hazards, the world's impoverished have accumulated an estimated $9.3 trillion in unregistered real estate.

3.2. PROPERTY BENEFITS TO INDIVIDUALS AND SOCIETY

Formal property owners have many advantages that enable and motivate them to act in ways that benefit society. Possession of real estate grants the sole authority to decide how best to utilise one's resources. Owners are encouraged to invest in and upgrade their properties by the security and independence from outside influence that come with property rights. The ability to transfer property, which makes it liquid and a source of cash, is another important property right. This implies that an owner can make improvements to a property with the knowledge that the value will increase when the property is sold. It also implies that a home can be mortgaged to finance a small business's first operations. Better information regarding an acquisition and the assurance that the deal will be upheld are also advantageous to the buyer. Last but not least, a clear property system saves an owner time and money by eliminating the need to establish usage rights and settle property-related conflicts. Market economies are fundamentally based on property. Contracts cannot be settled without formal property, and only transactions between people who know and trust one another personally can be accomplished with reliability. Strong property rights and the execution of contracts allow for interactions between strangers and over great distances. Documents proving property ownership also reveal who is in charge of what assets. In turn, these benefits increase options, promote specialization and economies of scale, and boost overall economic efficiency and production. The definition and upholding of fundamental property rights are also necessary for other aspects of contemporary market economies, such as joint stock companies and complex private capital markets. It is becoming more widely acknowledged that property has a significant influence on innovation in addition to increased rates of saving and investment and efficiency advantages in production and exchange. Since the industrial revolution, market economies have grown tremendously due to rates of innovation that have never been seen before, in addition to improvements in efficiency. In turn, in addition to their well-known function in fostering price rivalry, markets have sparked competition over invention. They accomplish this, in part, by encouraging and rewarding entrepreneurship. Through the introduction of new companies, technologies, goods, and business practices, entrepreneurs alter the economic landscape. Property allows business owners to experiment, sell their ideas, and make money that balances the risks and expenditures they make. Entrepreneurs will not take these risks if there is no chance of gaining a competitive edge. It is important to emphasize the difference between invention and innovation. Inventions might be the product of pure accident, research and experimentation, different economic systems, and diverse sources of funding and motivation. Entrepreneurs and market systems in general, on the other hand, are excellent at innovation, which includes commercializing innovations, modifying technologies for real-world applications, and making numerous, frequently small-scale improvements to corporate procedures and goods. Paper, porcelain, moveable type, gunpowder, the spinning wheel, a mechanical cotton gin, and ship technology are just a few of the many amazing inventions that came from mediaeval China. Regrettably, the government regularly confiscated private property and inventions, which discouraged creative, profitable behaviour. Government expropriation is probably the reason why these amazing technologies were not extensively used and did not spur growth of any historical significance. Stronger protection of intellectual property rights (IPRs) is needed in modern China, according to businesspeople, in order to maintain recent success and provide new opportunities for innovative companies. Although international businesses have voiced their dissatisfaction with the lack of effective protections against intellectual piracy, domestic producers have probably been more negatively impacted. Piracy, in particular, makes it difficult for major Chinese consumer goods manufacturers to build reputable brand identities. The reputation of the original inventors as well as their marketing expenditure are jeopardized by the inferiority of replica items. Higher-quality products for consumers and chances for producers to advance up the value chain are both made possible by an improved IPR framework. When paired with a wide variety of competitive marketplaces, an IPR system fosters the greatest social value in terms of innovation and technology dissemination. China has previously ratified international IPR accords, but its administration and enforcement still need work.

3.3. PROPERTY AND INSTITUTIONAL DEVELOPMENT

Even though property is often cited as the primary driver of growth, the rule of law is necessary for it to work. In other words, the government must create and uphold property rights, and a legal system must equitably enforce them. Jeremy Bentham made the compelling argument that "property and law are born and must die together." There was no property prior to the laws; if the laws are repealed, all property ends.11. In this sense, property cannot be established separately from justice and sound governance. On the other hand, there is a strong need for the growth of governmental and legal institutions due to the general desire for property and the financial advantages it offers. To put it another way, property reform can promote the growth of several significant institutions. "Property can be thought of as a tree, with social institutions as branches that grow from the trunk it produces." Property is a basic institution that fundamentally alters people's behaviour. In other words, property (in conjunction with competition) transforms a shared goal for material gain into actions that benefit society. Adam Smith's theory of the invisible hand is essentially this. As a result, entrepreneurs are urged to take chances and think outside the box to enhance output and customer service. To paraphrase

Lawrence H. Summers, "No one has ever washed a rented car in the history of the world." Owners, on the other hand, are expected to take responsibility for their belongings and handle them carefully. Everyone gains a stake in society when they are given the chance to possess some property. Furthermore, stable property rights might deter rent-seeking, which is a practice when producers look to the government for resources, protection, and privileges in an effort to avoid competition. The government is burdened by this behaviour, which also diverts resources from their most productive and efficient purposes. At its worst, it can stifle individual initiative and promote an unhealthy reliance on the state. Establishing safe property rights can help producers focus on productive and creative endeavour rather than the government and their rent-seeking interests. Fair competition results when numerous private producers have safe property rights and government control is constrained. When forced to uphold property rights, sovereigns also behave differently. In the past, sovereigns rarely or never distinguished between their private property and public lands and income. They had the power to expropriate or redistribute private property and riches at will. The development of contemporary property systems, which essentially followed the development of constitutionalism, brought about this transition. Civil freedoms could no longer be violated by governments. They were held responsible by taxpayers since they had to follow and abide by the law. The right allocation and use of public and private properties would now be determined by taxpayers and property owners. Since then, political and economic liberties have both grown and been destroyed globally, but property has been acknowledged as a fundamental right in Article 17 of the Universal Declaration of Human Rights. There are four ways that property supports democracy. It offers defense for the individual's civil liberties and safety. Second, it promotes competition and pluralism while reducing state control over economic resources. Third, constitutionalism and the rule of law are necessary and related to it. Lastly, it enables affluence and possibilities that increase democratic legitimacy and give people a stake in democracy.

3.4. TRANSITION TO FORMAL PROPERTY SYSTEMS

Property is an exclusive right by definition. The majority of people are now included in the formal economy, and in fact, the global economy, thanks to the expansion of this right beyond the elite class. The establishment of a just and efficient property system opens up new opportunities and removes obstacles to development in developing and transitional nations. It takes much more than just coming up with a thorough, uniform set of rules and laws to create such a system. Laws must be supported by the appropriate level of administration and enforcement, and they must be in line with regional traditions and economic realities. For example, when state-owned businesses were privatized in Russia, outside specialists were consulted to create a cutting-edge corporate law. Perverse repercussions resulted from the privatization since this law and others were not enforced. While the rights of minority shareholders were disregarded, the assets of the privatized firms were seized by the new controlling owners, who exploited their political ties. Due to the requirement that they lease their premises from the government, small company owners in Russia were also denied protected property rights. Corruption grew in this setting and hampered the growth of a separate, respectable private sector. The property rules that are already in effect in many emerging nations were draughted with the viewpoint of those who are already employed in the formal economy in mind. As a result, the majority of economic activity in the informal sector is ignored by these rules. Expansion of rules and strict enforcement typically fall short of their intended goals in these nations. More economic activity is forced underground as a result, and it becomes more difficult for unofficial business owners to enter the formal sector and earn a living.

The issue of definin□sufficient property ri□hts for the informal majority was ele□antly resolved by Hernando de Soto in Indonesia. In the rice fields, he found that a different dog would bark every time he stepped over an unseen property line. Therefore, one may map the existing property boundaries by listening to the barking dogs. According to De Soto, local customs are established by individuals worldwide to meet their real estate needs. Their holdings can be transformed into assets with worth that goes beyond the local community by lookin□into these customs and incorporatin□them into official law. Laws based on regional customs and norms are more practical and legitimate, which makes them more widely accepted and upheld (De Soto, 2007). By providing deeds of incorporation for enterprises and titles for residences, it has actually been able to increase legal ownership on a wide scale. 6.3 million impoverished Peruvians have been able to legally own their real estate since the 1980s, according to the Institute for Liberty and Democracy. Additionally, according to ILD, the formalization of 380,000 enterprises increased yearly tax collection by $300 million and created 560,000 new employment. Families are investing more on their homes as a result of entitlement. In Buenos Aires, at least 40% of newly entitled households have comparatively high-quality roofs and walls. Additionally, these families made greater investments in their kids' education. After obtaining title, residents of urban squatter settlements in Peru made 68% more investments in home improvement. However, having a title hasn't always made it easier to obtain bank credit, which frequently requires revenue from a professional job. It should be noted that property rights, not credit, are a major factor in deciding the amount of money that entrepreneurs invest, even though these findings imply that mortgages might not become the main source of funding for business owners in developing nations. It should be obvious that the poor gain from the formalization of enterprises and homes because these assets are more secure and reliable. New and important issues are also brought up by this formalization. To enable property exchange, contracts, credit extension, bankruptcy resolution, and efficient enforcement, more institutional advancements and modifications will be needed. The development of a truly market economy is made possible by this set of institutions and liberalization. However, providing fundamental property rights is a great place to start and is definitely required.

3.5. PROGRAMS TO STRENGTHEN OWNERSHIP AND REDUCE INFORMALITY

An subsidiary of the US Chamber of Commerce, the Centre for International Private Enterprise (CIPE) supports private sector groups worldwide that promote improved business conditions and more robust property rights. CIPE supports the institutional underpinnings of both free-market economies and democratic political systems by drawing attention to the reform of property rights. The delineation of property rights through titling and business formation is just one aspect of CIPE's programs. Complementary institutions are necessary for the defense and enjoyment of these rights. For example, a business's shareholders would not profit from a law of incorporation if onerous licensing requirements prevented the business from operating as intended. Regulatory hurdles, government harassment, corruption, ineffective courts, and poor corporate governance can all diminish the value of ownership for businesses. By addressing these different institutional shortcomings, CIPE aims to promote entrepreneurial growth, improve ownership, and lessen informality. Based on CIPE's experience, the most effective way to tackle these intricate issues is to engage both formal and informal business leaders in determining reform objectives and rallying them around solutions that benefit the business community as a whole. A coordinated lobbying effort attracts legislators' attention and increases reform momentum. Three of the many campaigns that CIPE has assisted private sector leaders in launching to address property-related concerns are detailed below. These initiatives have been similar in that they have prioritized local control of the reform process and lobbying.

3.5.1. Russia

Access to nonresidential building stock was a challenge for small and medium-sized businesses in many Russian regions. Despite the privatization of many companies and buildings, regional or local governments nevertheless controlled the land on which they were located in order to continue receiving rental income. High rents, short lease terms (usually one year with no option to buy or extend), and a lack of motivation for lessees to engage in property upkeep and repairs were all factors. For instance, one in three entrepreneurs in the Primers region stated that the most costly and time-consuming activity they encountered was renting real estate. The property market became opaque and very harmful to economic activity due to steep rises in rental rates, a lack of effective procedures for calculating rent, an accounting system for repairs, and unfair contract practices. In order to create a number of business agendas for better private sector growth, CIPE convened business associations and influential people from eight different regions between 2003 and 2005. Colleagues from their area collaborated with the participants to determine shared challenges and important development goals. After that, they established strategic alliances to deal with local public figures on certain concerns. For instance, the Krasnodar coalition presented a model agreement on the renting of municipal property that included clauses covering the cost of capital repairs and giving renters in good standing the opportunity to extend their rental agreements. Additionally, the group suggested the creation of an easily accessible municipal property registry and competitive, economically sound rental rates for municipal property. Entrepreneurs received technical support on lease documentation from specialists. As a result of the coalition's efforts, the municipal lease rate factors were revised to lower them for some entrepreneurs, including small trading enterprises. In another instance, the Primorsk group was successful in its efforts to stop a 250% increase in rental prices in Vladivostok on behalf of small businesses.

3.5.2. Montenegro

The Centre for Entrepreneurship and Economic Development (CEED), a CIPE partner and economic think tank, conducted an analysis of Montenegro's business registration procedure. This investigation revealed the need to streamline the registration procedure, lower the needed capital, and make it less expensive and time-consuming. Through a complex web of organisations and at the whim of government bureaucrats, CEED determined that registering a firm takes at least 45 working days. Businesses also have to re-register annually. CEED effectively pushed for reforms through the working group tasked with draughting the new Enterprise Law and extensively disseminated its paper, "Barriers to Doing Business in Montenegro," among the business sector and legislators. The 2005 Templeton Freedom Prize for free market solutions to poverty was to CEED for "Barriers to Doing Business." CEED was heavily involved in the draughting of the Enterprise Law, which was enacted in January

2002. Previously, it took $5,000 to register a firm; under the new law, it now takes four days and one dollar. If a company is not contacted by the government within a day, it is automatically reregistered. More than 15,000 business owners and 14,500 additional legal entities have registered as of right now. CEED provided free registration assistance to more than 4,000 businesses.

3.5.3. Kenya

Millions of small business owners and employees in Kenya's jua kali sector use primarily manual labour and minimal capital to supply goods to local markets; they frequently rely on handcrafted tools. These small enterprises are constantly harassed by local authorities, who make arbitrary arrests, chase them out of their temporary premises, seize their goods and equipment, destroy their property, demand bribes, and occasionally even attack the business owners. Business owners find it difficult to obtain loans since they are unable to obtain ownership of their stores and properties. The expansion of jua-kali enterprises is further hampered by onerous laws and regulations. Jua-kali enterprises have united and established their own associations in order to better their operating circumstances and safeguard themselves from exploitation by the government. A number of these associations have benefited from the abilities and tactics their leaders learnt during a CIPE training in Nairobi in 2004. Since then, they have raised their organizations' level of professionalism and boosted membership perks. Operators of jua kali are typically required to seek multiple licenses for the same business and pay numerous taxes that they cannot afford. Nonetheless, the groups have secured affordable single licenses for their members. The National Informal Sector alliance (NISCO) is a new grassroots alliance formed by participating associations from the 2004 workshop. NISCO has pushed for the government to provide basic amenities like power and water, restore shed buildings that house Jua Kali businesses, and give members access to titled land. In addition to government pledges for funding and procurement possibilities, NISCO has secured a sizable Nairobi hawker's market, which is anticipated to be finished by the end of 2007. The market was inaugurated with a ceremony presided over by the Kenyan president.

3.5.4. From Informal to Formal: "Top Ten" Priorities

Participants in a workshop organised by CIPE at the World Movement for Democracy assembly in November 2000 came up with a list of 10 priorities for legalizing the unorganized sector and utilising its potential for expansion. These priorities lay out a series of coordinated actions to strengthen democratic government, enable owners to exercise their rights effectively, and establish property rights frameworks. Even though property rights aren't the only issue with informality, solving these issues requires an institutionalized, comprehensive strategy such as:

- Establishing inclusive, transparent decision-making processes and giving the public frequent chances to voice their opinions on new and current laws and regulations;
- Removing unnecessary, redundant legislation that raise transaction costs and encourage corruption in order to simplify and clarify legal and regulatory frameworks;
- Ensurin□that everyone's ri□hts to private property are well-defined and strictly upheld;
- Simplifying and lowering the cost of business licenses and permits to enable businesses with little resources to join the formal sector;
- Modifying tax structures to support widespread compliance and profitable economic activity;
- Modifyin□labour re□ulations to make them more flexible;
- Improving corporate governance laws to promote free markets as opposed to crony capitalist systems in order to reform economic systems. These include outlawing insider trading, revising bankruptcy laws, and implementing globally recognized accounting standards, disclosure □uidelines, and anti-trust and anti-conflict of interest measures;
- To ensure that laws and regulations are administered and enforced effectively, efficiently, and affordably, □overnment entities that are overly bureaucratic should be reformed and their administrative and enforcement capabilities strengthened;
- Providing all residents with access to business-related information, such as how to start a business, create a commercial entity, get a license or permission, and mana□e a firm; and
- Establishin□ the hi□h-quality, cost-effective, and efficient infrastructure required for a market-based, democratic society so that entrepreneurs can profit by enterin□the official sector.

Development organisations, scholars, and governments of developing nations are increasingly in agreement that building fundamental institutional reform requires collaboration with local change agents, particularly those in the private sector and civil society. Like many others who advocate for democratic and market-oriented changes, CIPE's experience in developing nations undoubtedly supports the notion that the private sector and civil society must be the main forces behind reform. They can offer useful information about how new institutions and regulations will impact them, as well as who can garner political backing for positive reforms. To create a well-run, growth-oriented environment, they must take charge of institutional reform. It has been demonstrated that the creation of a just and efficient system of property rights has wide-ranging, constructive effects on development. The formal institution of property has been a vital component of sustainable growth throughout history and in many regions of the world. It has also aided in the growth of additional good governance entities. However, no one property regime is universally applicable. Due to their complexity, property systems must be customized for the local environment rather than being implemented out of the blue. What is deemed acceptable is influenced by historical and cultural variables. In order for property laws to be effective, they must be in line with the court system as well as other pertinent laws and organisations. When creating a property system, it is important to remember that property needs to be dynamic rather than static. Property is more than just a treasure that can be kept in a vault like gold. Innovations and increases in production create new riches and better well-being, not gold. Property is a dynamic asset that makes it possible to generate fresh wealth and investments. In a similar vein, property is not merely taken or transferred from one person to another; rather, it is crucial to increasing the overall amount of wealth that is accessible in ways that are advantageous to society. Strong property rights legislation should not be seen as a temporary fix but rather as an essential element of long-term, sustainable growth. Other essential institutions, like the legal system, contract law, regulations that guarantee competition, and even representative political institutions, are necessary for a full, functional market system. Implementing property reform has expenses, such as resolving disputes and administrative charges. Nevertheless, overcoming these obstacles and putting in place a property system can aid in the development of a community. Although every civilization starts from a distinct place, respect for property and the encouragement of investment, trade, and creativity can be advantageous to all communities.

3.6. ECONOMIC DEVELOPMENT AND PROPERTY RIGHTS

The right of an owner to use a good or asset for consumption and/or revenue generation is known as a property right (sometimes known as "use rights"). Known as "transfer rights," it may also include the ability to give it to another person through a sale, gift, or bequest. By renting, pledging, or mortgaging a good or asset, or by permitting other parties to use it, as in an employment arrangement, a property right also usually confers the right to enter into a contract with other parties. The significance of property rights, sometimes known as the "relations of production," in the process of economic development was important to the classical economists, from Smith to Marx, but mainstream economics has only recently adopted this viewpoint. Economics' fundamental welfare findings about the function of competitive markets are predicated on the idea that property rights are clearly defined and easily upheld. Traditionally, the research on economic growth has concentrated on capital accumulation and savings in a world without institutions and with perfect property rights. However, in recognition of the need to explicitly depart from a frictionless environment, the new institutional approach to development economics (North, 1990) has placed issues about effective property rights at the forefront of development thought. As said by North: Institutions are a society's rules of the game, or more formally, the limitations that people have created to influence how people interact with one another. As a result, they shape human transaction incentives, be they social, political, or economic. According to this viewpoint, property rights are a crucial component of an economy's institutional framework. Property rights, on the other hand, are not endowed; rather, they develop throughout time as a result of political and economic processes. As a result, a study of property rights must also take into account the legal and informal agreements that guarantee the clarity and enforcement of property rights. The role of the state in establishing and defending these rights has gained more attention recently due to developments in political economy. Economists usually use the term "property rights" to refer to private property rights, one of which is the ability to legally bar others from utilising an item or good. This influences how resources are allocated by influencing people's incentives to engage in productive activities that involve using the good or asset, making investments that preserve or increase its value, and trading or leasing it for different purposes. Nonetheless, in many countries, other types of property rights—like community property rights—are significant. People have the right to utilise common property, like a lake or forest, but they do not have the authority to stop others from utilising it. Additionally, some assets have limited owner transfer rights. Slavery, for instance, is illegal in contemporary economies. Generally speaking, property rights (including transfer and use rights) are always limited; for instance, the owner of a piece of land does not have the authority to engage in unlawful activity there. The political, legal, and enforcement framework in effect at a given period and location determines the type of constraints that are imposed.

When these concepts are examined, it soon becomes evident that property ri□hts have many si□nificant aspects that are fundamental to the functionin□of economies and provide incentives for people and businesses to make profitable investments. The phrase "effective property rights" refers to several aspects that are important to the economy. It primarily refers to the fact that ownership structures—whether individual or collective—are clearly specified. This affects consumption and wealth distribution in a first-order way. By determinin□ who has the ability to use an item and enabling the separation of ownership from control, property rights also have an impact on the pattern of production. Thus, the evolution of property rights determines the size and makeup of rental markets. By influencin□ whether assets can be passed from parents to children, these rights also have an effect on how wealth distribution changes across generations. To the extent that assets can be pledged against default, rights may also have an impact on how markets, especially credit markets, evolve. The connection between contracts and property ri□hts is a si□nificant conceptual topic. Both outline a range of decision-making powers, such as the ability to forbid certain behaviours or to take others. Since these rights can be outlined for every conceivable scenario, a rental agreement is essentially the same as a change in ownership in a world with perfect contracting. The renowned Coase theorem (Coase, 1960) is based on this principle: resource allocation will be unaffected by property rights allocation in a society with perfect information and no contracting costs. Since not all uses of a good or asset can be predetermined for every scenario, owning and renting are not equal in a world where contracting is expensive. The notion that property rights grant the owner residual control rights is a corollary of this. For those who possess them, these rights are a source of freedom that gives them the ability to choose how they want to use the thing (subject to any legal or technical limits). The incentives of the holder to make investments to increase the asset's value as well as those of other parties who may have contractual rights to utilise the asset would be impacted by this.

3.6.1. Resource Allocation and Property Rights

By restricting expropriation and promoting market exchanges, clearly defined property rights essentially enhance resource distribution and boost investment in a particular community. Indeed, only when property rights make engaging in socially beneficial activities worthwhile can there be economic progress (North and Thomas, 1973). Therefore, the existence of weak property rights is seen as a major barrier to economic growth (*see, for example*, Johnson, McMillan, and Woodruff, 2002; North and Thomas, 1973). In a society without institutions and with perfect property rights, the literature on economic growth has historically concentrated on capital accumulation and technical advancement. Building on the groundbreaking work of Coase (1960), the new institutional approach to development economics (*see, for example,* North, 1990) has now acknowledged that property rights are expensive to create, define, and enforce, and as a result, they will never be perfect (*see, in particular,* Barzel, 1989). As a result, the institutional approach to development now acknowledges that the provision of property rights has a substantial impact on resource allocation in a world with positive transaction costs, which can be thought of as the resources used to create and preserve property rights. The Coase theorem is undoubtedly one of the most well-known ideas in economics. It asserts that resource allocation is effective regardless of ownership patterns when rights are well defined and transaction costs are zero (Coase, 1960).One However, the system of property rights that a society adopts will have an impact on how its economy functions because, in fact, the costs of transactions are positive and the costs of defining and upholding rights are high. In fact, as Coase has highlighted, his goal was to demonstrate the essential role that transaction costs do—and ought to—play in forming the institutions that comprise the economic system, not to imply that a world with zero transaction costs was a realistic depiction of reality (Coase, 1988). Following in the footsteps of Coase (1960), Barzel (1989) highlights that property is vulnerable to appropriation by others due to the difficulty of precisely measuring all of an asset's features. As a result, rights are never completely defined. Property rights guarantee enforcement, which promotes specialization. However, for the same reason, expensive institutions and resources are needed to preserve property ri□hts after they are transferred (Arrun ada, 2012).

In the process of economic development, land ri□hts are especially si□nificant. The primary production asset for agricultural activities is undoubtedly land. Furthermore, it lends itself to use as wealth and collateral due to its immobility and near indestructibility (Binswanger and Rosenzweig, 1986). However, land ri□hts are still □enerally weakly defined for historical, economic, and political reasons. In fact, out of all the competing and excludable assets, land is most likely the one for which ri□hts are not clearly defined, especially in developin□ nations. It is important to examine the precise nature of such expenses in further detail because the absence of a clear definition implies a possible loss of societal efficiency. Land rights have the potential to impact economic performance in a number of ways. First, if people are afraid that others may take advantage of their investments, they may underinvest due to insecure land rights (Demsetz 1967; Alchian and Demsetz, 1973). The fact that millions of urban dwellers lack official titles to the land parcels they occupy is a widespread example of weak property ri□hts in the developin□ world of today (Banerjee and Duflo, 2007). Second, barriers to parcel transferability may indirectly impact the distribution of resources, leadin□ to economic inefficiencies (Besley, 1995). Third, if there is no land titling, there may be less chance of land collateralization, which restricts access to finance. Finally, uncertain land ri□hts may have si□nificant long-term implications on total productivity and poverty reduction by influencin□ decisions about labour, fertility, and health. Both urban and rural settings may experience these effects differently, and wealth levels may play a si□nificant role. All of these ar□uments lend credence to the idea that a lack of secure property rights to land may restrict opportunities for development, investment, and credit access. For this reason, many bilateral and multilateral institutions have sponsored programs that either register land titles or enhance the operations of land administration institutions. For instance, the World Bank alone was overseeing a portfolio of land administration projects totaling over $1 billion USD in 2005. The literature has addressed the effects of various land rights systems in great detail. However, because property rights are usually endogenous, it is challenging to determine the impacts of land property rights. We review the literature in this study and look at a selection of studies that try to pinpoint the causes of land property rights.

3.6.2. Land Rights, Expropriation Risks, and Transaction Costs

The definition of property rights is closely related to the consideration of transaction costs (Barzel, 1989). Assets like land that are legally owned grant their bearer a number of rights. In its fullest sense, ownership grants the right to use the item, to keep others from using it, to give the asset to someone else, and to keep these rights throughout. Gains from trade will be reduced if these criteria are partially absent. People will use resources to try to protect their ownership rights to their assets rather than seeking profits that benefit both parties, which will raise transaction costs (Barzel, 1989). Economic examination of property rights focusses on how changes to these rights affect the distribution of resources. The Problem of Social Cost, written by Coase in 1960, was a revolutionary addition to this body of work. Coase started by thinking about the significant historical case of cattle that roamed the streets. He specifically examined the situation of a rancher whose land bordered a nearby farm. Occasionally, the farmer's crops were trampled by the rancher's cattle that wandered into his property. How does production change based on who owns the land? was the question raised by Coase. The profession was taken aback by Coase's response. He came to the conclusion that, regardless of the assignment of liability, the farmer and the rancher could bargain and resource allocation would be the same if the transaction cost was low enough.3. The Coase theorem is the name given to this astounding realization. The most crucial thing to keep in mind, though, is that the Coase theorem only applies when there are no transaction costs and property rights are clearly defined. The distribution of property rights does affect how resources are allocated when transaction costs are positive. Property rights, however, are not a binary issue. From having no rights at all to having a full range of rights, there is a spectrum. In fact, people still have economic rights over assets even if they do not have legal rights. These rights are equivalent to the power to use an asset's services directly or indirectly through exchange (Barzel, 1989). We can define property rights in accordance with the extent of the group executing such rights and the level of control granted to the exercising group, as per Schlager and Ostrom (1992). According to this definition, property rights can be separated into two categories: private and collective. The latter can then be further separated into open and government property. It is crucial to stress that the extent of common-pool losses, the type of contracting costs required to settle such losses, and the financial costs associated with establishing and upholding property rights will all influence whether a more fully defined set of property rights—that is, formal private property rights—will be advantageous to society (*see, for instance,* Ostrom, 1990).

However, there are other standards and guidelines that govern land rights in practice besides formal land rights. Customary law is still quite powerful and applies differently in each nation (Plateau 2000, for example). Land rights also have multiple dimensions. While alternative methods of definin□property rights, such public titling, may be more effective in fostering trade, security of tenure, for instance, may offer an efficient means of encoura□in□investment. Apart from the availability of legal rights, developing nations also differ greatly in how these rights are enforced. Legal land rights could not have much of an impact in reality if there are no institutions in place to properly enforce property rights. In fact, North (1990) contends that the primary reason of underdevelopment is societies' incapacity to create efficient, affordable means of contract enforcement. Effective judicial systems in modern nations have clearly defined le□al frameworks and a□ents, and citizens have a reasonable amount of faith that the merits of a case—rather than personal gain—will determine its conclusion. On the other hand, enforcement in developing nations is unclear due to both uncertainties in legal doctrine and uncertainty about the court system's conduct. The state protects legal rights in contemporary cultures, but it does not always decide the extent of those ri□hts based on maximisin□ social benefit, subject to transaction costs (*see, for example*, Libecap 1989). On the other hand, poor governance is typically found in emerging nations (Baland, Moene, and Robinson, 2010). Land property rights for the poor or disenfranchised groups in society are frequently compromised as a result of elites' state takeover (*see, among others,* Binswanger, Deininger, and Feder, 1995; Baland and Robinson, 2008). Therefore, there is room for the promotion of the adoption of land policies that aim to formalize legal rights as a means of enhancing welfare and lowering poverty, given that the provision of legal rights is likely to be suboptimal, particularly in developing nations and for the impoverished population. The high cost of learning about the legal system and enforcing rights frequently makes the drawbacks of a weak institutional structure worse (Deininger and Feder, 2009b). Therefore, the institutional framework in which land registration or land titling occurs, the effectiveness of the state in upholding legal rights, and the effectiveness of the enforcement of informal land rights all have a si□nificant impact on the value of these processes. Si□nificant variation in the consequences of legal rights results from the likelihood that different societies and people within a society would experience varying transaction costs while enforcing property rights. Lastly, the dilemma of how to proceed remains even when land titles are to be provided. The relative effectiveness of the two most si□nificant titlin□ systems—re□istration and recordin□—is examined by Arrun ada and Garoupa (2005). Accordin□ to their findin□s, re□istration seems to be more efficient, at least in industrialised nations, and the efficiency level rises with the number of land transactions.

3.6.3. Effects of Secure Land Property Rights

First and foremost, protected land rights increase resource allocation efficiency by reducing expropriation risk and encouraging investment. People will underinvest if others can profit from their investments, as the literature has stressed (Demsetz, 1967; Alchian and Demsetz, 1973). Because they make land holdings more transferable, secure land rights—particularly the holding of public titles—can also attract investment. Investments in nameless land would be extremely illiquid even if the risk of expropriation were extremely low, but titling lowers the cost of alienating the relevant assets (Besley, 1995). First and foremost, in agricultural settings, investments and land improvements should result in increased land productivity and income. On the other hand, it is unrealistic to anticipate that housing improvements in metropolitan areas will directly impact earnings. Second, by promoting market transactions, protected land rights improve the effectiveness of resource distribution. Higher-valued land can end up in the hands of producers and consumers thanks to transferability. Renting is also made easier by protected property rights. Because they might not be able to reclaim their land, untitled landowners could be reluctant to rent it out. Without leasing opportunities, economies of scale may be lost in agricultural areas. Furthermore, while trying to engage in the nonrural economy, landowners could encounter obstacles (*see* Deininger and Feder, 2009b; and the references therein). If rental activity is hindered in metropolitan areas, migration and appropriate household-house matching may be hindered. Collateralization of assets is a third way that legal land rights can influence investment and resource allocation. Many have suggested that the use of land as collateral in credit markets is hindered by the absence of formal land titles (Feder et al., 1988; De Soto, 2000). Effective property rights can make it easier to employ assets to reduce agency costs when they are there. Possessing legitimate land rights may help debtors pledge their land holdings as collateral in the credit market, where enforcement costs are high. Besley and Ghatak (2010) examine the use of land as collateral using a straightforward formal model that includes moral hazard and restricted liability. They make the assumption that each borrower has a certain amount of money and that as property rights become more lawful, a larger percentage of that wealth will be available as collateral. They show that marginal increases in collateral security have no effect on resource allocation in the credit market if the borrower's wealth is either low or high; this effect is limited to an intermediate range of wealth endowments. According to this intriguing theoretical realization, no credit consequences should be anticipated if the legalized land assets have very modest values. This finding implies that attention should also be paid to heterogeneous impacts that may arise depending on the distribution of wealth and other characteristics of the population, but it does not invalidate the search for the identification of average effects of titling programs in empirical work. It is obvious that how the credit markets function has a significant impact on how titling affects credit access. Credit could increase investment and profits in rural areas, but this potential effect is negligible in developing nations due to the near-complete absence of formal credit markets (Deininger and Feder, 2009a). Even while private lending institutions are more active in urban credit markets and they are generally more established than rural ones, the urban poor nevertheless have limited access to formal credit (*see, for example*, Banerjee and Duflo, 2007). Purchases of homes are frequently financed by residential credit through collateralized long-term loans, and these transactions require proof of land ownership. However, it is rare and probably limited for the impoverished to use residential real estate to support entrepreneurial ventures. From the perspective of loan demand, the impoverished might not have any worthwhile initiatives or might believe that their own home is too valuable to risk in an entrepreneurial endeavour. On the supply side, banks may find it challenging to reclaim substandard homes in underprivileged areas due to political and fixed legal expenses, and the proceeds from their sale may be rather small (*see* Deininger and Feder, 2009b).

Therefore, whether land registration or land-titling programs are large enough to cause scale effects will also likely affect the impacts of land titling. Think about the credit market, for instance. In a location where only a tiny percentage of impoverished people own land titles, it might not be advantageous for financial institutions to extend credit. On the other hand, if a sizable portion of the populace obtains legal land ownership, this could alter the incentives for financial institutions, encoura□in□business entry and, consequently, expandin□ the poor's access to credit. Lastly, the structure and distribution of resources within households may be impacted by the presence of secure land rights. For instance, landowners may have to invest money to protect their rights if their legal rights are not fully developed. By reducing the resources required to uphold property rights, land rights can promote efficiency in households with limited labour resources (Lanjouw and Levy, 2002; Field, 2007). However, enhancing land ri□hts is unlikely to have a si□nificant impact on labour supply if families are not limited in terms of their overall labour supply (Besley and Ghatak, 2010). However, this influence may vary □reatly across rural and urban environments. Furthermore, the lack of official land ri□hts may prevent impoverished families from having a useful savings and insurance tool that could shield them in retirement and in difficult times, forcin□ them to rely on their children and/or extended family members as insurance mechanisms (Galiani and Schargrodsky, 2011). Once more, these impacts might be different in rural and urban settings. For instance, cultivating several plots in rural locations may give households some insurance, which lessens the value of acquiring individual property rights (Platteau, 2000). Titling status may also have an impact on household size and, consequently, investments in the human capital of children through these mechanisms.

3.6.4. Endogenous Property Rights

Why do some nations become wealthier while others stay impoverished? The real per capita GDP of the US, Germany, and Japan is about 15 times higher than that of nations like India and Pakistan. A number of economists have recently made the case that a nation's economic growth is significantly influenced by the calibre of its institutional framework (*e.g.,* North and Thomas, 1973; North, 1987, 1990; Clague et al., 1997). North (1990) asserts that the primary cause of both historical stagnation and the current underdevelopment of third world nations is the absence of institutions that define and defend property rights. To put it simply, economic progress necessitates incentives for capital accumulation, which in turn necessitate solid claims to the products of accumulation (Grossman and Kim, 1995). The protection of property rights and economic growth are strongly linked, according to a number of studies, including Kormendi and Meguire (1985), Scully (1988), Barro (1991), Torstensson (1994), Keefer and Knack (1995), Hall and Jones (1999), and Clague et al. (1999). These works are based on the economic theory of predation. The instability of property rights caused by predation—extortion, appropriation, theft, and litigation—delays the creation of capital, according to Hall and Jones (1999). However, there aren't many theoretical studies that integrate the theories of economic growth with predation.

Assuming a basic growth model with overlapping generations, in which both property rights protection and predation are determined endogenously, an attempt is made to explain the relationship between property rights protection and economic growth. The difficulties of both individuals and businesses are impacted by the protection of property rights. While businesses pick the portion of capital allotted to protection, individuals choose the portion of labour allotted to predation. If we incorporate honesty into both the utility function of individuals and the profit function of businesses, we can account for the consequences of social institutions. It is a factor that reduces the appeal of labour devoted to predation in the utility function. It is a factor that increases the effectiveness of capital allocated to the protection of property rights in the profit function. The model produces a distinct yet dynamically inefficient equilibrium in the absence of governance. The level of output is rising in this equilibrium in terms of the discount rate, the rate of utility of honesty, and the rate of protection efficacy. In addition to forecasting output levels, the model also forecasts welfare levels. If the discount rate, the protection effectiveness rate, and the honesty utility rate are all high, we discover that the output equilibrium level is higher than the golden rule level. The model produces either a unique equilibrium or numerous equilibria when the government is included as the exclusive supplier of protection as a public good funded by a capital tax levied on businesses. In the situation of multiple equilibria, there is a low equilibrium with low output, low protection, and high predation, or a predation trap, and a high equilibrium with high output, low protection, and low predation. Which equilibrium the economy converges to depends on the initial level of output. The rate of protection, or the percentage of output that the government can safeguard, is likewise low if the initial level of output is low, as this will result in low tax revenues to support protection. People are hence highly motivated to appropriate. Because of this, people devote a little percentage of their time to labour, which pushes the economy into the predation trap, a low equilibrium. The rate of protection, or the percentage of output that the government can safeguard, is high if, on the other hand, the starting level of output is high. This is because substantial tax revenues are also created to pay protection. There aren't many incentives for anyone to appropriate in this situation. The economy reaches a high equilibrium as a result of people allocating a large portion of their time to work.

Some of the previously described components can be found in a number of current research, including Grossman and Kim (1995), Tornell (1997), Zak (2002), and Gradstein (2003). A dynamic general equilibrium model with endogenous predation and property rights protection is put out by Grossman and Kim (1995). In addition to dividing their initial endowment between production and consumption, two groups of people also choose to divide it between protection and predation. One of the most significant findings from their model is that, despite having perfectly protected property, the generations of the group of people who opt to avoid predation build up capital more slowly than the generations who came before them. Deterrence is an expensive option, even if it turns out to be preferable to accepting predation (Grossman and Kim, 1995). Tornell (1997) offers a model of economic growth that incorporates endogenous property rights protection, much like Grossman and Kim (1995). He examines how property rights change between common and private property regimes as the economy grows. Every individual has access to total capital under common property. Every person is only able to access their own capital stock under private property. Individuals' predatory actions to gain access to a bigger portion of the capital pool are what cause regime transitions. He makes the assumption that the common property re□ime will prevail at first, and he demonstrates how the economy can become stuck in this regime and suffer from slow growth perpetually as a result. However, if the economy expands and the economy shifts to private property, it becomes worth it for people to pay for the establishment of institutions that safeguard property rights. But as the market grows, predation becomes more alluring, which encourages people to undermine the institutions and return the economy to a system of common property. A distinct property rights regime, where only a select few have access to the total capital pool, is the subject of Zak (2002). The concept put forth by Tornell (1997) suggests that predation is unidirectional. Predation occurs between those who have access to the aggregate capital stock and those who do not. According to Zak, nations can become impoverished, and even if they manage to break free, people with shaky property rights will still have lower incomes for the rest of their lives. Furthermore, developing nations may still be unable to break free from the poverty trap even if predation encourages the endogenous development of institutions to safeguard property rights and even if such institutions optimally allocate resources to this protection. Gradstein (2003) suggests a model that also generates several equilibria, one with high property rights protection and a high income level, and another with low property rights protection and a low income level, assuming that property rights protection is a public benefit. Due to the hi□h cost of property ri□hts protection, only wealthy nations will be able to invest the funds necessary for protection, which will spur economic expansion. There will be a greater desire to defend property rights as the economy expands.

One of the main concepts is that both the development of productive capital and the protection of property are inherent to the process of economic expansion. A general-equilibrium model of the interaction between two dynasties—one of which is a potential predator and the other is its prey—serves as the analytical framework. Every generation of every dynasty must choose how to divide its endowment of inherited riches between defensive fortifications and offensive weapons in addition to consumption and productive capital, as in conventional growth models. Wealth accumulation is based on productive capital, but predation has the potential to both destroy and redistribute wealth from the victim dynasty to the predator dynasty in each generation. According to this concept, the prey dynasty's current generation has the option to devote a sizable portion of its inherited wealth constructing defensive structures in order to ward off predators. As an alternative, the prey dynasty's present generation may decide to put up with predation and devote a small portion of its inherited wealth on defensive fortifications in order to lessen its impacts. The present □eneration of the prey dynasty decides to accept predation rather than try to prevent it if the potential predator dynasty's wealth is low in comparison to the prey dynasty's wealth. Additionally, the predator dynasty's inherited wealth increases over generations in comparison to the prey dynasty's inherited riches. Consequently, the rate at which the prey dynasty's productive capital accumulates and the security of its property both gradually decline. As the predator dynasty's relative wealth rises, it allocates more resources to offensive weapons than the prey dynasty does to defensive fortifications, which reduces the security of the prey dynasty's property. Two factors cause the prey dynasty's productive capital to accumulate at a slower rate: First, the prey dynasty devotes a growing portion of its inherited money to defensive fortifications and a declinin□portion to productive capital as a result of the predator dynasty's □rowin□relative affluence. Second, each succeeding generation of the prey dynasty inherits a lesser portion of the productive capital of the generation before it because to the waning security of the dynasty's property. A generation of the prey dynasty will eventually discover that, as the predator dynasty's wealth increases in comparison to that of the prey dynasty, its property would become so insecure if it allowed predation that it would be better off enhancin□ its defensive fortifications to the point where they discourage predation. The prey dynasty's fortune is completely guaranteed from this generation on. Only for generations of the prey dynasty that tolerate predation do the rate of capital accumulation and property security have a positive relationship. Even if their property is completely safe, generations of the prey dynasty that chose to discourage predation build up productive capital more slowly than their predecessors who allowed predation. Therefore, in contrast to popular belief, slower capital accumulation may be linked to increased property security. Because deterrence necessitates a si□nificant investment of inherited money in defensive fortifications, this outcome is achieved. Deterrence is an expensive option, even if it turns out to be preferable for the prey than accepting predation

Chapter 4

Property Rights in Agriculture

Contents

OVERVIEW

At the national level, property rights are important because they guarantee both individual and national prosperity, but they are also important for specific industries in each nation. Agriculture is an example of a sector that has a centuries-old connection to property rights. Private property rights, both through historic resource ownership and more recent forms of ownership over technology breakthroughs, are the primary source of support for the agricultural sector. Farmers and ranchers need recognized property rights, either through title or lease, to effectively regulate agricultural resource use. Control is crucial for ownership of livestock and reliable access to valuable resources like land. When secure access or ownership is established, farmers and ranchers are more willing to take risks and experiment with new methods. This maximizes resource utilization and boosts overall productivity. Land is one of the fundamental needs for cultivating crops or rearing livestock, thus if ownership and control over resources were lacking, there probably wouldn't be any agricultural sector at all. Farmers and ranchers have a variety of rights under property laws that they can use to decide how best to produce their goods. "The ability to bar others from using the resource is a key component of a property right." The Supreme Court established this specific right in Kaiser-Aetna v. United States, wherein the Court prohibited the Army Corps of Engineers from proceeding with pond improvements that were privately owned because doing so would have resulted in the loss of "one of the most essential sticks in the bundle of rights... the right to exclude others." Further, the Court ruled that the government could not grant the public a right of access to privately owned property. Deterring infringements of private property rights, such as trespassing, by requiring the government to provide fair compensation, promoting trust in the property rights system, and discouraging property owners from using self-help remedies when they believe their property is in danger were the goals of rulings like Kaiser-Aetna. Courts like the Supreme Court of New Jersey have affirmed and stated a few exceptions to the right to exclude. However, despite these exceptions, the right to exclude is still effective, as shown by the classification of the right to excludes significance in cases like Kaiser Aetna, Hodel, and Jacque, where courts—including the US Supreme Court—have favored the right to exclude over a conflicting interest. "It appears that this only considers the absolute right [to exclude]." The ability to keep others off your land so you can utilise it exclusively is a fundamental right. According to renowned law scholar and property law author scholar Richard Epstein, "it is obvious that the right to exclude is part of [the bundle of rights]" and that this right "must be protected." Additionally, state and federal courts have put a lot of effort into defending this right against infringement by the public or the government, and they have consistently acknowledged it inside the bundle of sticks. Because property law and the courts have established a pattern of protection, owners have become accustomed to the idea that their pleasure and ownership of their property will not be disrupted.

In the agricultural sector, it is particularly crucial to have the freedom to exclude others and utilise one's property however one sees appropriate, provided that one does not infringe upon the rights of others. Land ownership is the primary source of income for farmers and ranchers. Land is particularly valuable since it is a scarce resource and is necessary for the production of crops and livestock. The ability to keep people off one's property gives farmers and ranchers peace of mind that their precious resources will be protected, which encourages them to take chances like trying out novel techniques and boosts overall productivity. Courts have maintained that private property rights and the ri□ht to exclude others come first, even if the individual violates someone else's property rights under the pretense of another specific ri□ht. In Robins v. Pruneyard Shopping Centre, for instance, the court determined that the shopping center's authority to bar anyone from entering does not infringe upon the First Amendment's safeguards of free expression. This decision demonstrates the California Supreme Court's conclusion that the right to bar people from private property does not infringe upon the rights of free speech when it is necessary for the use or economic worth of the property. Since a farmer or rancher depends on their property for their livelihood, their rights over it are crucial to its use and economic worth. All things considered, property rights—particularly the ability to bar others from your private property—are crucial to the agricultural sector. Property rights that are safe and upheld promote the required risk-taking behaviours in the sector, including development and the adoption of novel techniques.

Property rights and land tenure have an impact on how technologies are used in agriculture and natural resource management. Protected property rights provide farmers with enough incentives to boost their productivity and maintain environmental sustainability. It makes sense that in the absence of protected property rights, farmers would not invest in land development, use resources effectively, or feel an emotional connection to the land they cultivate. The United Nations Centre for Human Settlements (UNCHS) asserts that property rights and land tenure are closely related. Secure property rights to land are primarily justified by the incentives they offer for sustainable □rowth and land investment. According to Deininger (2003), there are several ways in which property rights impact economic growth. First of all, protected property rights will encourage more investment by households and people and frequently will □ive them better access to finance, which will not only allow them make such investments but also serve as a backup plan in case of shocks. Second, it has long been understood that the operational distribution of land influences output in unmechanized agriculture; hence, a very unequal allocation of land will result in lower productivity. Secure and well-defined land rights are essential for household asset ownership, productive development, and factor market functioning, even while the ability to use land productively will depend on policies in areas outside of land policy that may require independent attention.

4.1. LAND TENURE

Government laws and agricultural policies, which covertly affect farmers' profit margins and land use decisions, regulate the security of land tenure and market prices of agricultural commodities (Ehui et al., 1994). In particular, one of the key factors influencing farmers' decisions on land use is their land tenure (Feder and Nishio, 1998; Pandit and Thapa, 2003; Rasul et al., 2004). Farmers require significant investments in land development in regions where shifting cropping is prevalent. However, if they are unsure of how long they will be permitted to exercise the ownership right, it could influence their investment choices. Tenant farmers in particular are hesitant to invest in land management if they do not obtain land tenure rights, which leaves them open to eviction by the government or landlords. Farmers are more concerned with short-term profit maximization at the expense of hastening land deterioration if property rights are lacking and land tenancy is unstable (Feder and Nishio, 1998). Farmers who lack land ownership or have insecure land tenure are also less able to obtain the credit necessary for better farming practices (Fede et al., 1988; Rasul, 2003). Even if they are willing to change, their inability to obtain financing drives them to stick with old land-use practices (Thapa, 1998b; Rasul and Thapa, 2003; Rasul, 2003). Accordingly, national policies impact institutional arrangements including finance and marketing facilities as well as infrastructure development, which in turn affect land-use systems (Bergeron and Pender, 1999; Rasul, 2004). Although land reform has varied greatly by geographic region, land tenure institutions have always taken agricultural and economic growth into account (Maxwell and Wiebe, 1998). Through the dismantling of large estates and the transfer of land resources among the rural populace, land reform in Latin America meant changes in the size of land holdings. In Africa, land tenure reform usually refers to the legal changes in land tenure that are meant to improve security of tenure with the goal of increasing productivity and promoting better land conservation practices. In East Asian contexts, land reform meant "land to the tiller" or the dissolution of landlord/tenant relations. In a nutshell, the land reforms focus on improved conservation practices, increased equity, productivity, and food security through tenure adjustments. According to Thiesenhusen (1995), land tenure reform had the following effects:

- Food security;
- Reduction in social polarity;
- Increased investment;
- Transparent production incentives;
- Poverty reduction;
- Increased employment; and
- Greater equity.

Figure 4.1 illustrates the connection between agricultural productivity and land tenure (either through institutional reforms or legal changes).

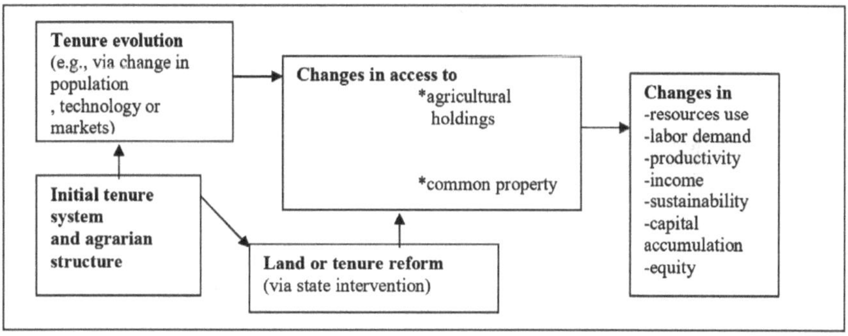

Figure 4.1. *Conventional characterizations of tenure's impact on production and income.*

Source: Maxwell and Wiebe (1998).

The FAO defines land tenure as the connection between individuals or groups with regard to land, whether legally or conventionally defined. (For ease of use, "land" is used to refer to other natural resources like trees and water.) Land tenure is an institution, meaning that communities have created norms to control behaviour. The distribution of land ownership rights in societies is governed by rules of tenure. Along with related obligations and limitations, they specify how access to the rights to use, control, and transfer land is granted. To put it simply, land tenure regimes establish the conditions and duration of use of resources. Land tenure is frequently classified as follows (*for more information, see Chapter 2)*:

- **Private:** the transfer of rights to a private party, which could be a business entity like a non-profit or commercial organisation, a group of persons, a married couple, or an individual. For instance, individual families might be the sole owners of residential, agricultural, and specific tree parcels within a community. Without the permission of the people who own the rights, other community members may not be allowed to use these resources.

- **Communal:** A community may have a right of commons if all of its members are free to use its resources as they see fit. Community members might be entitled to graze cattle on a shared pasture, for instance.
- **Open Access:** No one has any particular rights, and no one may be excluded. This usually includes maritime tenure, where anybody can access the high seas; it can also include rangelands, woods, etc., where everyone may have unrestricted access to the resources. (One si□nificant distinction between communal and open access systems is that the former prohibits non-community members from using the common spaces).
- **State:** In the public sector, certain authorities are given property rights. For instance, in certain nations, whether at a central or decentralized level of governance, forestlands may be subject to governmental authority.

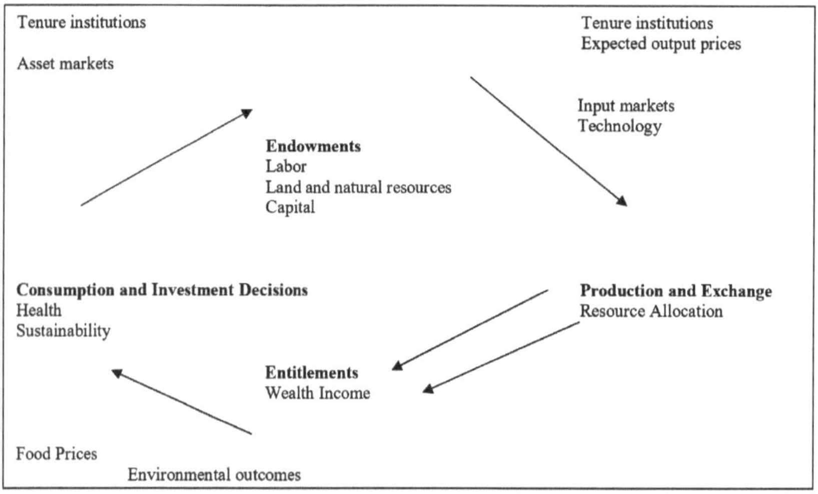

Figure 4.2. *Land tenure and food security.*

Source: Maxwell and Wiebe (1998).

Land tenure structures in agrarian societies have a direct impact on household food access by limiting access to resource endowments, and also have an indirect impact on regional or national food security through food availability (and consequently, food costs). One of the key factors influencin□ access to food is resource availability. According to Maxwell and Wiebe (1998), there are three key aspects of food access: sufficiency, sustainability, and vulnerability. It can be seen as a function of food availability, access, and utilization. Decisions about production, marketing, consumption, and investment lead to structural changes in the distribution of resources across families, demonstrating the dynamic relationship between land tenure and food security. Figure 4.2 shows the dynamic interlinkages:

4.1.1. Property Rights

Instead of focusing on who "owns" land, property rights theory examines the legal and unwritten rules that establish who is entitled to benefit streams that result from the use of assets and who is not (Bromley 1991, Eggertsson 1990, Libecap 1989). For these rights to be legitimate claims, they must be approved by a collective. Therefore, in order to support a claim, property rights entail a relationship between the right holder, other people, and a governance system. The rule and the means by which it is enforced are the two parts that make up property rights. State legislation, customary law, user group norms, and other frameworks may serve as the basis for the regulations. Since the state is typically in charge of enforcing statutory law, the rights are based on formal legislation. Customary authorities or a user group that controls the allocation of rights or whose members "define or enforce rights among themselves" can both enforce property rights based on various kinds of regulations (Schlager and Ostrom, 1992). A broad classification along the following property regimes is frequently used in the recent literature on property rights over land and other natural resources: open access (no rights defined), public (held by the state), common (held by a community or group of users), and private (held by individuals or "legal individuals" like companies). According to BendaBeckmann and Benda-Beckmann (1999), such a classification can only provide as a rough reference to the actual effective entitlements that a right holder in one of the stylized property regimes holds.

- **Bundles of Rights:** Schlager and Ostrom (1992), whose focus was mainly on examining common property regimes used by collectivities of resource users, developed the concept of bundles of rights, which is helpful for examining the distribution of rights among various user groups and between individuals under various tenure regimes in Ethiopia. Their classification of bundles of ri□hts is thou□ht to be equally useful, nevertheless, when analyzing various sorts of land tenure regimes, particularly when identifying hybrid regimes that are not entirely private, common property, or governmental ownership. As a result, it is thought that examining the varying degrees of rights granted to Ethiopian rural peasants over time, space, and political regimes provides a compelling illustration of this type of analysis.

4.1.2. Property Rights and Agricultural Productivity

The primary institutional asset that rural residents rely on to support themselves is their property rights to land. In fact, the best indicator of poverty in many nations is landlessness. Farmers' willingness and capacity to accept investments and inputs that increase productivity are significantly impacted by the nature of their property rights to land. Farmers in the industrialised world typically have strong and secure land rights, but this is not the case for a large portion of the developing world. Secure land rights are those that are socially and legally legitimate, well-defined, enforceable, long-lasting, and suitably transferable. Regretfully, many smallholders in developing nations lack at least one of these essential elements of robust land rights. Even worse are women farmers. Furthermore, landless farm workers—the most marginalized households in the agricultural sector—face even more difficulties. The rural poor frequently have limited alternatives for utilising land to enhance their standard of living in the absence of strong land rights. Thankfully, there are alternatives that have been shown to increase welfare and productivity. Furthermore, a large number of international development organisations are in a good position to help and encourage those solutions.

4.1.3. Land Rights and Smallholder Vulnerability

Smallholder farmers make up a sizable section of the global impoverished, and the nature of their land property rights is closely related to both their production and poverty: The majority of the impoverished in rural areas have some access to land, despite the fact that many of them lack it. Not only do these "landed poor" have small holdings, but they also have weak and insecure land rights, which keeps them impoverished. Their motivation to invest in their land over the long run or use it responsibly is weakened by the uncertainty they face. Since their land cannot be lawfully transferred, its economic value is limited. Some people believe that all "owners" have explicit rights to the property they utilise, and that all smallholders cultivate the land they own. Systems of land rights, however, are dynamic, intricate, and predicated on various kinds of property interests. These rights are not only about having solid possession; they are a continuum. While some smallholders have only limited rights to land that the government considers to be publicly held, many smallholders are actually tenants who farm land that belongs to others. Unfortunately, the need of knowing who has what rights to the land involved in the program is often overlooked when implementing government and donor projects. Because of this, these schemes frequently unintentionally weaken smallholders' tenure security and run the danger of making them even impoverished. Smallholders with the least amount of power are typically the most at risk. One example is the worldwide land rush in developing nations. Although there is a lot of potential for increasing agricultural output by introducing much-needed finance, technology, and better business models, the rush seems to be displacing many smallholders who do not have secure rights at the moment. In actuality, smallholders with shaky property rights may be even more vulnerable to losing their land as agricultural land becomes more productive and, thus, more appealing to those with means and influence. If the poor's land property rights are not protected, expanding a resource's agricultural potential is likely to result in elite appropriation.

4.2. BENEFITS OF LAND REGISTRATION AND TITLING

Since the beginning of time, most societies have been preoccupied with land and the arrangements surrounding its use and possession. The patriarch Abraham, who owned a lot of livestock but no property, needed to buy a piece of land some 3700 years ago so that he may bury his late wife. Abraham proposed that he sell the land to Efron the Hittite after he found a suitable location (which featured a cave, which was a highly valued burial place in those days). Abraham refused Efron's offer of free use of the land and insisted on paying for it instead. Apparently because he wished to prove his (and his heirs') undeniable possession by paying money. In addition, many citizens witnessed the transfer of possession as the actual payment was made at the city gate, a customary public gathering spot, in front of the city's notables (Genesis, 23). The requirement to confirm who owns the land being sold in order to eliminate the possibility of future disputes that may cost the buyer money or even result in the loss of property is the reason why witnesses and publicity are given such importance in today's real estate transactions. Although some of the risk is mitigated by conducting transactions in the public domain, this structure is inefficient, particularly when the volume of transactions increases. The aforementioned biblical allusions attest to the fact that the need for land registration has arisen in many societies as land-use technology advancements, population growth, and increased trade have necessitated the documentation and enforcement of property rights on land (Binswanger et al., 1995).

In their efforts to advance economically, many emerging nations view land registration as a top priority. As they work to become more market-based, transitional economies have shown a special interest in land registration. Since they must divide their limited resources among various uses according to the anticipated advanta□es, decision-makers in developin□ nations would benefit from havin□ a comprehensive understandin□ of the scope and type of benefits that may be realistically anticipated from land registration. In order to make unbiased decisions on land registration investments, it would be very helpful to have answers to some basic questions re□ardin□the economic benefits. What increases in agricultural output, for example, would a nation receive from land registration, and how would these gains compare to the advantages of, say, extension services? Under the correct conditions, land registration systems can have si□nificant economic and social effects, as will be shown in this chapter's empirical section. As such, they ought to be regarded as a major area of public investment.

4.2.1. Economics of Land Registration

The impact of titles on improving tenure security and their function in collateral arrangements, which make it easier to obtain institutional credit, are the two sources of the relationship between titles and economic performance that have been noted. Complementary investments in drainage, structures, tree and stone removal, and other enhancements are necessary for the productivity of land in all of its uses, whether it be agricultural or urban/commercial. By definition, the cost and labour of these investments are borne up front, but they pay off over time. Thus, expectations about the duration of the horizon over which the investor may gain have a substantial impact on the incentive to make these investments. These expectations, in turn, rely on the likelihood that the investor's ownership of the land will be contested, evicted, or expropriated by the government. The term "tenure insecurity" is frequently used to describe these problems. Throughout human history, land titles and registration systems have developed as an institutional framework to lessen this kind of instability. The risk of ownership issues is decreased, the possibility of having to pay hefty fees to defend one's land ownership is decreased, incentives to invest are improved, and land productivity is raised when ownership is formally recorded and validated. Since the lender promises to refund the money in full over time, with interest, credit transactions are by their very nature dangerous. Collateral arrangements, in which fixed assets, usually real estate, are used as a guarantee for loan repayment, have emerged because lenders usually have less knowledge than borrowers about the likelihood that the loan will be returned in full and on time. When borrowing from formal lending organisations, who frequently have incomplete information about the borrower, a secure title is required to mortgage land since the lender must be convinced that the borrower is, in fact, the owner in order for the land to be used as collateral. Although debtors without a secure title can obtain credit from informal lenders without utilising land as collateral, these loans are normally only available for short periods of time, are typically significantly more expensive than official credit, and are frequently of tiny amounts. Therefore, a farmer without a secure title is limited in their ability to increase output and productivity. Therefore, the elimination of this restriction may be a significant source of financial gain from land registration. In reality, according to farmers in the Thai study mentioned above, the primary advantage of having secure titles was the ability to obtain financing.

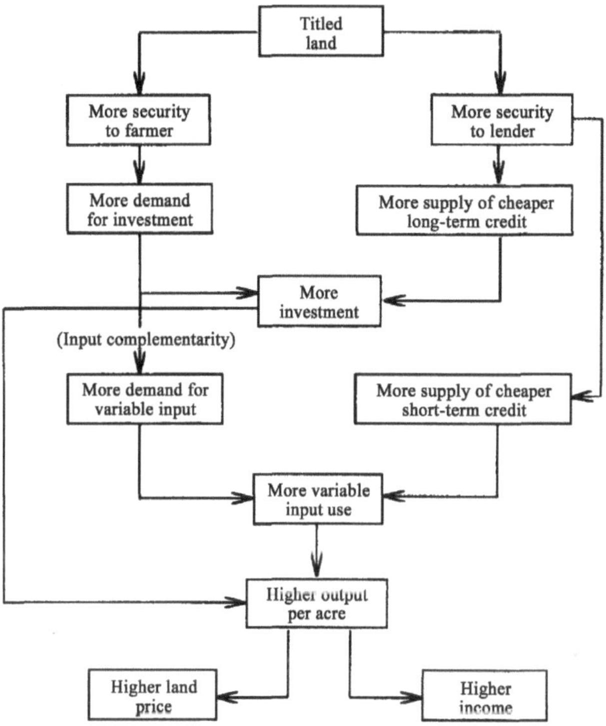

Figure 4.3. *Land ownership security and farm productivity.*

Source: Feder, G., and Nishio, A. (1998).

The newly named farmer can now expand investment and input use, resulting in increased production per unit of land, by obtaining credit at a lower cost and larger amount. Since the price of land generally reflects the capitalized worth of the stream of incomes (or residential services) that it delivers, higher land productivity subsequently results in higher land prices per unit. There is another way that these systems assist society economically, even while the framework mentioned above justifies land registration and titling arrangements based on gains in production per unit of land. The distribution of the skills and resources needed to make the most economical use of land does not always align with the distribution of land ownership or possession at any particular time. Therefore, it's feasible that the land's productivity under its current owner is lower than it would be under a different operator who might be more skilled or have the resources to make complementing improvements. These short-term inefficiencies in land allocation are typically resolved by land sales and rentals, as those with potentially higher-value uses for the land are able to offer the current (low value) user a price (or rental income) that is higher than the land's capitalized value, resulting in an exchange of ownership or possession and, eventually, an increase in productivity (Figure 4.3).Although property transactions have existed since the be□innin□of time, their scope is influenced by the information asymmetry between buyers and sellers. As long as most transactions are between people who live in the same town, information is fairly symmetrical and the buyer may quickly determine whether the seller is legally permitted to sell. Conventional systems typically develop to manage community transactions and can work well. However, as land markets grow and more transactions take place between unrelated people, questions about an owner's right to transfer property rights become more pertinent. It is possible to improve market efficiency by reducin□ or eliminatin□ ownership ambiguity by giving landowners titles that are supported by a legal system that can enforce their property rights (Feder et al., 1988). When these agreements aren't in place, the uncertainty that prospective purchasers experience lowers the amount of demand for land and lowers the offer price. Registration systems and titling reduce uncertainty to a considerable extent, which leads to more transactions, better land distribution, and increased economic well-being. The aforementioned model merely offers a broad framework, and its applicability to a particular nation is contingent upon its policies, customs, culture, and other unique elements.

4.2.2. What Makes Land Registration Viable?

4.2.2.1. Economic Viability

Improved credit availability, increased land-attached investments, increased use of variable inputs, and more effective land markets are the primary drivers of increased revenue and better residential services. Thus, when land is utilised as collateral (probably through institutional channels), there should be reasonably well-functioning financial markets that are able to offer long-term credits. Land registration systems won't offer advantages connected to the credit market if different policies limit or prohibit the enforcement of collateral, or if the administration of collateral contracts' laws and enforcement is too onerous to be efficient. In terms of current economic opportunities, there should also be a clear incentive to increase output per unit area through the use of more variable inputs and land-attached investments. Farmers may not have much of an incentive if produce markets are too far away, land is infertile, agricultural input distribution systems are not operating efficiently, labour shortages persist, etc. For instance, it wasn't until Thailand opened up to international trade that the first significant investment in the country's land registration system in the late 19th century became feasible. This led to an increase in land cultivation as rice farming became extremely profitable. Similarly, when the overall economic climate is not favourable for business, urban investors might not be motivated to increase their investments, even if land ownership security is improved through registration. However, in cases where there are general incentives for productive investment, it is important to determine whether the current tenure system offers adequate tenure security (e.g., as demonstrated by the high frequency of land disputes). In many cases, a system of customary property ownership offers enough security to encourage investment. It is impossible to undervalue the advantages of land registration brought about by increased land market efficiency. If there is little demand for land transactions in a given area or if customary land rights systems are adequate to support the kinds and volumes of transactions that are normal in the area under consideration, the absence of a formal land registry won't restrict land markets. It is commonly known that a large number of customary tenure systems are effective at facilitating the transfer of land rights between individuals, primarily within a particular group of people. For example, even in some metropolitan areas, a large portion of the 5 million hectares of land in Indonesia's West Sumatra province are subject to a traditional system of communal rights (hak ulayatf). Over generations, the arrangement has been effective in facilitating matrilineal transfers of land rights. It is evident, therefore, that the tenure security offered by the conventional system is insufficient to defend the locals' claims against outsiders, whether they be squatters, developers, or investors in golf courses. There must be an enabling regulatory environment for land registration even in cases where there is a latent demand for land transactions that is limited by the lack of land register and title documents. The rules and legislation pertaining to land registration are one important area. Realistic and socially acceptable proof criteria, incentives to register (such as giving registered parcels precedence over unregistered parcels), and explicit dispute resolution methods should all be included in these laws and regulations. The absence of rigorous regulatory constraints that impact land transactions is another crucial point. Excessively complex land use approval processes typically result in higher transaction costs and less activity on the land market. There are a number of restrictions and prohibitions that limit land transactions and can negate the benefits that are anticipated from enhanced land market transactions under a formal land registration system. Some countries also set up land committees or bureaus that are tasked with ensuring that sale transactions do not violate social or cultural norms. It might not be the best time for the government to invest in land registration if it is unclear whether the essential elements for the systems' economic sustainability are there. Other forms of investment, which address more enduring barriers to the society's economic growth, would be more beneficial.

4.2.2.2. Social Aspects of Land Registration Programs

Those who are better informed, more accustomed to formal procedures, and have greater access to officials and financial resources may be able to "land grab" opportunities when a modern registration system is introduced to replace a traditional (and usually less formal) system. Claiming ownership of land that was previously owned or used by others who are less knowledgeable or lack the resources to complete the registration processes, or even asserting exclusive rights to land that was previously state, communal, or open access land, are examples of land grabbing. Wealthy people purchasing untitled or unregistered land at a discount with the understanding (not known to the original owners) that land registration will soon be possible at a relatively cheap cost is a considerably less sinister but still unfair scenario. Policymakers often argue that smallholders and the poor will be enticed to sell their land to larger landowners or to those with more money in the resulting more active market, which is why they oppose formalizing land rights and lowering the cost of land market transactions. It is argued that this will cause societal instability and the rise of a sizable class of people without land. One could imagine situations where asymmetry of information and opportunities could result in land sales that are socially undesirable, even though the theoretical foundations of these arguments are not always sound. Anecdotal evidence sometimes cites cases of abuses of the type mentioned above in a variety of nations, often made worse by authorities who, in the name of political expediency or greed, collude with more powerful and wealthy people. When titling and registration are irregular (that is, done on demand by individual landholders) and when the cost of registration is considerable, there is a greater chance of such unfavorable results. On the other hand, formal registration systems can have highly positive equality implications when informal systems are under pressure from shifting economic conditions. This is because they give the poor a low-cost way to defend themselves against attacks to their informal rights. A government can follow a number of guidelines to lessen the possibility that land registration programs will have an unfavorable social impact such as:

- **Transparency:** Transparency should be the primary guiding principle in the administration of the land registry. While some governments restrict access to the land registry, usually in an effort to curb land speculation, doing so would encourage abuses by those with access, create uncertainty in land transactions, and diminish the accuracy of land records. The public should be informed about the administrative processes and associated costs for land registration using a variety of media.

- **Stakeholder Involvement:** Second, and in keeping with the previous point, landowners ought to be actively participating in the registration procedure. Unless the prospective beneficiaries and their neighbors are required to present proof and documentation, ideally on-site with neighbors, and are asked to clarify boundaries, submit testimonials regarding other residents, etc., land registration can be challenging and unfair. To increase landholder involvement and the likelihood of fair outcomes, strategies such as beneficiary consultation, reliance on communities and community-based organisations, ease of use, and speed of delivery can be employed.
- **Program Evaluation:** Third, before launching a land registration program, important societal factors that may impact the advantages of land registration should be considered. The effects on women and rules pertaining to customary land tenure are two areas that would require extra attention. If current laws make it impossible for women to benefit from land registration or perhaps negatively impact their rights, then land registration would not be advantageous for them. For example, the requirement that land only be registered in men's names may have such a negative effect. Before deciding whether or not to include customary lands in the land registration program, it is crucial that society (including the government) understands the nature of the current customary rights and the potential effects of registration. Until provisions are developed to enable benefits to customary right holders, customary areas should not be included in the land registration program if the potential impact on customary rights is uncertain or detrimental.
- **Effective Monitoring:** Fourth, there ought to be efficient tracking of local developments. According to preliminary findings from fieldwork conducted in regions that are systematically registered, there have been no reports of land-grabbing or other incidences in Indonesia that have had a significant detrimental social impact. On the other hand, the data showed that the registration effort was widely covered, especially by the impoverished, and that there was comparatively little rent seeking.
- **Cost:** Fifth, in order to maximize net social benefits and enhance sustainability, cost effectiveness should be a key consideration when

creating a land registration system. As much as possible should be done with already-existing tools, such as cadastral maps, informal ownership proof, and base maps. By taking use of economies of scale, Thailand's systematic registration approach—as opposed to its sporadic one—has helped to keep registration costs low. Wider and more equitable dissemination of information about registration opportunities and procedures is made possible by systematic registration. When selecting technology for tasks like mapping, control densification, and land records administration, cost should also be a key consideration. Even more affordable choices might be available thanks to recent technical advancements like handheld global positioning devices. The current human resource base, the executing agency's institutional capability, and the possible returns from other land-related investments should all be taken into consideration before makin□ si□nificant investments in complex land records management systems. In order to encourage low-income landholders to register their properties, the initial registration fee should also be maintained modest. Thailand's approach, which subsidizes first-time re□istration and recovers costs throu□h registration fees on subsequent transfers, has proven effective and is being extensively researched.

Althou□h there mi□ht be other standards, many nations would find it difficult to uphold the aforementioned crucial ones. In many instances, this would entail si□nificant adjustments to the current administration and operation of land registration.

4.3. FARM PROTECTION

State statutes known as "farm protection laws," or "ag-gag laws," make specific activities on farms illegal in an effort to thwart undercover investigations into the agriculture sector. "Agricultural interference" statutes were the first ag-gag legislation that states enacted. The tendency of prohibiting recordings and photos at private farming facilities without the owner's permission was started by these regulations. Alongside the interference laws, rules referred to as "agricultural fraud" legislation were also created.82 The use of false pretenses or misrepresentations to enter or secure employment at an agricultural production plant was prohibited by this type of ag-gag regulation. The majority of the ag-gag laws that the states proposed in 2011 and 2012 combined the ideas of "agricultural fraud" and "agricultural interference." However, the absence of reporting requirements made it difficult for these laws to pass the legislatures; as a result, towards the end of 2012, lawmakers draughted a new kind of farm protection law. "This new type of regulation, called "rapid reporting" legislation, requires anyone who takes a picture or makes a sound at an industrialised farming operation to give the recording to the authorities within a certain period of time, typically between 24 and 48 hours. Idaho has one instance of a farm protection statute. In 2014, in response to specific incidents at one of Idaho's dairy farms, the state passed its farm protection statute. In 2012, one animal rights activist went undercover to work at the dairy in Idaho. This activist then surreptitiously recorded some of what he perceived to be animal mistreatment, which the animal rights organisation Mercy for Animals later made public as a depiction of animal maltreatment. The dairy workers were shown on camera thrashing and kicking a dead cow. The public outcry over the video prompted the dairy's owner to take action, and the employees implicated were let go. Threats were still made against the owner and his family, nevertheless. As a result of these circumstances, Idaho passed its farm protection, or ag-gag, statute in February 2014. A person who enters an agricultural facility under false pretenses, obtains a facility's records, obtains employment at an agricultural facility, or purposefully causes damage at an agricultural facility without the owner's knowledge or consent is guilty of interfering with agricultural production under this statute.

4.3.1. Need For Farm Protection

State lawmakers enact agricultural protection legislation to shield ranches and farms from interference with their production methods. Any sorts of actions, including trespassing and vandalism, could cause interference. However, the agricultural business is extremely concerned about undercover examinations of farming operations by some activist groups; these so-called "undercover investigations" have the potential to cause serious financial harm to the farm (damaged property, reputation, lawsuits, etc.). The number of farms and ranches is steadily declining as smaller organisations are forced out due to pressures, such as growing costs and regulations brought on by these economic injuries. Only larger farms and ranches can grow and increase production to offset rising expenses, which reduces the overall number of farms and increases nonfarm employment. This could have an impact on the general public by reducing the variety of manufacturers and goods available in the market. Interferences can affect the owners of small farms in addition to perhaps raising expenses, which reduces the variety of products available to society by driving out small farms. 99% of farms and ranches in the United States are currently family-owned, meaning that many people rely on the agricultural sector for their livelihoods. Therefore, lives, including those of the owner families, would suffer when interferences, such undercover investigations, occur in a farming production facility, causing financial harm or perhaps the plant's closure. State lawmakers sought to enact farm protection legislation in order to prevent adverse effects because of the magnitude of the influence that interference with agricultural facilities had on the lives of farmers and consumers. For instance, according to the lawyer for the Idaho Dairyman Association, the goal of Idaho law is "to protect Idaho farmers from wrongful interference." The lives and labour of Idaho farmers are dispersed throughout the country, making them particularly susceptible to influence from improper behaviour. For farmers to be protected from vulnerability, they must have that assurance and consistency in their property rights. According to the law, undercover activists who "expose the industry to the 'court of public opinion,' which destroys farmers' reputations, results in death threats, and causes loss of customers" are another threat to farmers and the agricultural sector. In order to try to expose these claimed terrible practices, some citizen activists sneak into a facility—either by actual trespassing or by deceiving others—because they believe that farms and ranches treat their animals cruelly.105 There is probably little possibility that these so-called "investigative actions" will reveal any unethical behaviour that could be occurring. There is a much higher likelihood that they will find methods that appear strange or aggressive to people who have not worked in the agricultural sector but are actually entirely acceptable as industry standards. The strangeness or aggression may be used by these activist "investigators" to film the procedures in a way that makes them unrepresentative of what truly occurs on American farms. Additionally, the movies turn into a scare tactic for the general public, which hurts the farm and occasionally causes it to close permanently. Numerous parties may be impacted by this outcome, including but not limited to the farm's operators and their families, the workers and their families, the State, which would lose money if the farm closed, the citizens, who would no longer have a food producer, and all other economic activities and transactions related to the loss of a working farm. Although it would have been unheard of at the beginning of the century, covert investigations by citizen activist groups inside agricultural producing facilities are a reality today. These unnecessary investigations are the result of increased efforts and the rise in activist groups who, understandably, feel compelled to address the public's growing concerns about the origins and preparation methods of food by breaking into a farm or ranch and disclosing anything they can, whether it is good or bad.

People for the Ethical Treatment of Animals (PETA), Americans for Medical Advancement (AFMA), Animal Liberation Front (ALF), Antivivisection Coalition (AVC), Cruelty Free International (CFI), 11th Hour for Animals, For Life on Earth (FLOE), and The Humane Society of the United States (HSUS) are some of the larger organisations that are active throughout the United States. Although these groups occasionally put an end to improper activities occurring on farms and ranches, their reputation has been built on their history of abusing farmers and using whatever resources they come across—whether legal or not—to harm their property, exploit, and threaten them. Animal activists, for instance, attacked Ben Falloon, a poultry farmer, and Taranaki Farms in April 2018. Falloon raises his chickens in a shelter at night and lets them wander freely during the day, which is a more contemporary approach to chicken farming that many believe is more advantageous. For the sake of transparency and education, Falloon is quite transparent about his farming methods on social media. However, rather of applauding his efforts, a group of vegan animal activists started attacking him online. Falloon then thinks that this activist group broke into his farm in the middle of the night in April 2017 and attempted to steal—or, as he described it, "liberate"—the chickens from their cage. "They released more than 200 hens from the protection of their electric fence enclosure; dogs and foxes then savaged the birds throughout the night, and all of the released birds died." In addition to failing to "liberate" the hens by killing them all, the activists caused Falloon's tiny family farm to sustain damages totaling almost $40,000. Like many others, Falloon kept his livestock using modern farming methods, but the activist group did not understand his methods and chose to take matters into their own hands instead of interacting with him in a civil manner. This destroyed Fallon's property (the chickens) and did nothing to advance the activists' cause. Falloon's story is not unique; there have been numerous such examples in recent years of animal activist organisations going to considerable lengths to further their own agendas at the expense of farmers while disguising their actions under the guise of animal protection. PETA's actions are to blame for a number of these events. One of the most active animal rights organisations in the US, PETA has 6.5 million members worldwide, making it the largest animal rights organisation. Using public education, cruelty investigations, research, animal rescue, legislation, special events, celebrity involvement, and protest campaigns, PETA aims to address the four industries where the greatest number of animals endure the most severe suffering for the longest periods of time: laboratories, the food industry, the clothing industry, and the entertainment industry. Although PETA may have begun as a respectable organisation that engaged in legal activities to advance animal welfare, many believe that in recent years, they have veered away from their original goal at the expense of farmers and ranchers in order to further their own adapting agenda. across $40 million is donated to this organisation annually from all across the world. PETA could be helpful in reducing the numerous cases of animal maltreatment that take place worldwide, including poaching, threats to biodiversity, and overfishin□of oceans. "PETA isn't interested in that," thou□h. Rather, they are drawn to incredibly strange and nonsensical publicity stunts. Although these stunts aid PETA in gaining support and a following because the general public tends to view them as accurate representations of reality, they are typically founded on misinformation that harms livelihoods, promotes the spread of misinformation, and actually undermines the sincere advancement of animal rights. The most plausible explanation for ag-gag laws is to shield farmers and ranchers from the dissemination of misleading information that could seriously harm their enterprises, in addition to the infringement of their property rights. For instance, PETA members have worked undercover at dairy farms to enter the land without authorization, record the farmers' activities, and exploit them. In 2013, one member obtained a job under false pretenses and went undercover at Adirondack Farms in New York. Videos of the dairy staff frequently electroshocking downed cows, striking them with poles or sticks, or yelling were collected by this member. This may appear to be animal maltreatment to someone unfamiliar with dairy farms or cattle operations, which is why PETA utilised its undercover footage to target the issue. However, when it comes to working with cows, these techniques are rather widespread and occasionally even a necessary requirement, something PETA neglected to investigate or disclose to the public.

4.3.2. Impact of Invalidating Farm Protection Laws

Ag-gag rules will probably have a very detrimental effect on the agricultural sector as a result of the judges' rulings. This is due to the fact that private property rights will suffer if farm protection laws are declared unlawful, and the agricultural sector depends on secure property rights. Even while this impact might not seem like much at first, it might have repercussions in many other parts of the world.

4.3.2.1. Farmer's Decisions and Willingness to Take Risks

Farm protection legislation gave farmers peace of mind about their private property rights in the face of the current threat posed by covert animal activists and those who would exploit them. Farmers are likely to lose faith in their property rights and the long-term stability of those fundamental rights if that security is not in place. In many ways, this will influence the farmers' choices and willingness to take chances on their farm. First, clearly defined property rights are the source of investment, and the protection of those rights serves as an inducement to make an investment. An investor may be more inclined to fund riskier endeavour if their rights are more secure. Farmers can pledge property as collateral when security is in place, which encourages more labour and capital investment. On the other hand, if property rights are not solid and secure, "it may be difficult for businesses and individuals to reap the rewards of their investment and hard work." Second, the farmer can use resources and assets more freely and flexibly when they have secure property rights. Thus, a farmer can trade property as a factor of production or transfer assets to more productive users, giving them greater options. Increased efficiency and improved factor productivity are the outcomes of the greater mobility and flexibility that come with protected property rights. Third, farmers will invest resources in autonomous property protection, which is a waste of resources, if they fear that their property rights will be violated due to the lack of protection. These resources are free to invest in new methods and engage in productive activities on the farm or ranch thanks to protected rights. In general, a farmer is better able to obtain loans and is more inclined to invest in novel and risky endeavour when they have faith in solid and secure property rights. Furthermore, farmers who have protected property rights will probably be more flexible with their resources and use them more efficiently overall. They will be able to invest and use their resources more productively by putting less of them into protection. Protected property rights can lead to increased economic development, productivity, and allocative efficiency. However, property rights are now less safeguarded as a result of the courts invalidating farm protection legislation, which has a negative impact on economic development, productivity, and allocative efficiency. Since private property rights are the cornerstone of growth, this outcome has a cascading effect. The farmer is the first to be impacted by unsecured property rights. Farmers' per capita income is lowered when they lack confidence in their rights, are reluctant to take chances and invest, and devote resources to pointless endeavour like property protection. Then, because it is not growing and progressing, the entire agricultural industry is affected. "A permanent and enforceable land rights system must evolve" and be safe in order to offer powerful incentives for improving the land and adopting measures to restore its fertility. The current limited resources cannot produce enough food to support the world's expanding population unless this business advances. Furthermore, the environmental sector is impacted since farmers are not making investments in the newest machinery, embracing new technology, or putting more ecologically friendly techniques into effect. Because there will be fewer innovations and developments, society will subsequently see a decrease in the quantity of diversified and high-quality food products.

4.3.2.2. Trespass and Misrepresentations

The court ruled that Idaho's farm protection law was unconstitutional in part because trespass laws already addressed the issues the bill was attempting to prevent. In Idaho, "entering or remaining on another person's real property without permission, knowing or having reason to know that his presence is not permitted, constitutes criminal trespass and is guilty of a misdemeanor." In addition to preventing trespassing, the Idaho farm protection statute shielded farmers from undercover individuals who would pose as someone else in order to enter a production site. Since someone theoretically enters with the owner's knowledge and consent, misrepresentation differs from open trespass. The property owner cannot file a trespass lawsuit under this rule since, in theory, animal activists are visiting the land with the owner's permission, even if they are doing so under false pretenses. Consequently, unless actual harm is done, farmers have little to no defense against others who pretend to be on their private property in order to violate their property rights. Farmers may then file a lawsuit against the undercover activist for slander or bodily harm. But the farmer has surely already suffered from either irreparable harm to their reputation or damage to their assets and property. In an effort to close this glaring loophole in trespass law, Idaho passed the Ag-Gag law, which made it illegal to use false representations to obtain access to private property, thus violating the property owner's rights. This way, farmers would not have to wait until property damage or defamation occurred, by which time it would be too late.

4.3.2.3. Costs and Benefits of Ag-Gag Laws

The advantage of ag-gag legislation is far greater than the drawbacks, even though there may be a cost because criminalising undercover operations could result in some minor instances of animal abuse going unpunished. Having ag-gag laws in place has the advantage of making farmers and ranchers feel safe and secure in their property rights, which enables them to produce more food of higher quality and quantity to feed the expanding population and encourages them to take more chances in areas that will benefit from improvement. The farmer, society, the global population, and the environment all greatly benefit from this. Ag-gag laws not only prevent farmers from taking risks and investing in innovative methods that would benefit agriculture and society, but they also shield farmers and their property from infringement. Only when the invasion will benefit others, when the policy generates social advantages greater than the costs to the property owners, when the losing owners receive full compensation for the interference, and when harm to others is demonstrated, may interference with property rights be taken into consideration. Here, undercover activists fabricate information to obtain entry before evidence of animal abuse is established, so when they infringe against the farmer's property rights, the harm is uncertain. Given the rarity of discovering animal abuse on a farm, it is extremely improbable that an animal activist's covert operation will assist others or yield enough societal benefits to offset the significant financial burden on farmers due to the grave risk to their livelihoods. Moreover, farmers do not receive compensation for this infringement on their rights. Since the advantages of farm protection legislation clearly outweigh the costs of a few possible instances of animal cruelty, interference with a farmer's property rights should be prohibited. Furthermore, there are other ways to address animal welfare, but there are no other ways to address misrepresenting oneself to obtain access, which violates a farmer's property rights. Additionally, once videos of what animal activists believe to be unethical behaviour are released, the farmer's reputation is already damaged. Like the majority of states, Idaho has a statute that makes cruelty to animals a felony. According to Idaho law, "Anyone who treats an animal cruelly, causes or procures the cruel treatment of any animal, or who has custody or charge of any animal, whether as an owner or otherwise, subjects any animal to cruelty shall, upon conviction, be punished." People who break this rule risk having their rights revoked, paying fines of up to $9,000, and spending up to a year in jail. Therefore, addressing the ripple effect, safeguarding property owners generally, and upholding our long tradition of enforcing and defending individual property rights outweigh any potential minor costs associated with implementing farm protection laws. For the benefit of society, the economy, and the entire world, a few isolated instances of animal maltreatment should be ignored, and the agricultural sector should be given preferential treatment.

Chapter 5

Intellectual Property Rights

Contents

OVERVIEW

A product of human ingenuity and innovation, intellectual property (IP) is an intangible asset. IP could be a machine, a service, a formula, an invention, a product, or an idea. IP protects the legal rights of the original creator and prevents theft and misuse. The rights that protect an idea and the products or services derived from it are called intellectual property rights (IPRs). Although it appears to be a relatively new area of law, society has been using it for a very long time. In order to control the monopoly of commercial or noncommercial inventions and artistic works and prevent any exploitation of such ideas or inventions in the public domain, IPRs are territorial statutes that provide limited protection of intellectual material. IPRs are the foundation of commercial science because they offer legal certification for scientific concepts and ideas, which is crucial for commercializing a scientific discovery or idea. Despite being relatively new, IPRs are becoming a more interesting subject. Although it is still rarely taught in many scientific fields during regular coursework, it is a crucial skill for researchers, entrepreneurs, and businesspeople in the 21st century. Patents and copyrights have long been contentious issues among economists, but property and ownership are not.

On the other hand, trademarks are not contentious because they are only used to identify people and companies. A range of private contractual arrangements that are used to enforce "intellectual property," including as non-disclosure agreements, no-compete contract terms, and software shrink-wrap agreements, are similarly subject to the economic analysis of patents and copyrights. The debate over copyright and patents is important from a theoretical and policy standpoint. Because inventive activity is central to models of economic growth, trade, and industrial regulation, the theoretical relevance emerges. Advocates of intellectual property laws frequently state that their primary goal is to promote the production of a diverse range of intellectual commodities. In order to accomplish this, the law grants individuals and companies property rights to certain data and intellectual products they produce, typically for a brief time. Proponents contend that because intellectual property laws protect original ideas and stop ille□al copyin□, creators have more financial incentives to produce information and intellectual commodities in the first place because they can profit more personally from them. IP proponents think that these financial rewards and le□al safe□uards encoura□e creativity and advance specific types of technolo□y. As scientific and technolo□ical advancements challen□e their fundamental tenets, legal frameworks that grant innovators a temporary monopoly on the use of their inventions—most notably patents and copyrights—have grown more troublesome.

5.1. INNOVATION CONCEPT

Innovation is the process that leads to the first successful commercial use of an invention, discovery, or insight into a new system, technique, or product. As a result, it can be applied to new markets, new supply sources, new organizational structures, and new goods, procedures, and services. Despite their essential differences, innovation and invention are frequently confused. When a novel concept for a good, method, or service is initially conceived, it is called invention. The process that leads from the initial invention to its first profitable commercialization, however, is known as innovation (Freeman, 1985). Innovation, then, is both the process of change and the result of that change. Innovations can be classified as either new to the world, new to a nation, or new to an organisation. They can range from gradual to radical. Although management researchers focus much of their attention on radical innovations, a crucial—and frequently disregarded—source of strategic distinction is the accumulation of small improvements as innovations spread and change.

Generally speaking, innovation differs from creativity in that it focusses on applying innovative concepts in a business context. Based on the literature, Amabile and Pratt (2016) make a distinction between innovation ("the successful implementation of creative ideas within an organisation") and creativity ("the production of novel and useful ideas by an individual or small group of individuals working together"). The economist Robert Solow was able to show that economic growth was composed of two parts in 1957. The increase in output, which includes wa□e labour and capital, could be the cause of the first component. Productivity was determined to be the second factor. Since then, instead of presuming that technological advancements and inventions lead to increases in productivity, economic historians have attempted to understand the innovation process itself. Joseph Schumpeter's (1942) work, which characterized the economic consequences of innovative processes as constructive destruction, was largely responsible for the emergence of the notion of innovation following World War II. Consistent neo-Schumpeterian scholars nowadays do not view innovation as an apolitical or neutral process. Instead, it is possible to view innovation as a socially produced process. As a result, the political and social environment in which innovation occurs influences how it is conceived. "Innovation today is best understood as innovation under capital," asserts Shannon Walsh. The appropriation of knowledge (e.g., through patenting), the pervasive practice of planned obsolescence (including lack of repairability by design), and the Jevons paradox—which characterizes the negative effects of eco-efficiency as ener□y-reducing effects tend to trigger mechanisms leading to energy-increasing effects—all serve to illustrate the current hegemonic purpose for innovation, which is capital valorization and profit maximization. A number of frameworks have been put up to define different kinds of innovation.

5.1.1. Sustaining Versus Disruptive Innovation

Clayton Christensen's approach distinguishes between disruptive and sustaining innovations. The enhancement of a product or service based on the recognized needs of existing clients is known as "sustaining innovation" (e.g., faster microprocessors, flat screen televisions). On the other hand, disruptive innovation is the process via which a new good or service (such as a transistor radio or a free crowdsourced encyclopedia) opens up a new market and eventually drives out more established rivals. Christensen asserts that disruptive ideas are essential to a company's long-term success. Disruptive technology frequently makes disruptive innovation possible. According to Marco Iansiti and Karim R. Lakhani, fundamental technology is capable of laying fresh groundwork for long-term global technological systems. As completely new business models develop over many years, foundational technology tends to affect business operating models. Waves of institutional and technological change that accelerate more slowly are caused by the innovation's gradual and steady adoption. A foundational technology is the packet-switched communication protocol TCP/IP, which was first introduced in 1972 to support a single use case for electronic communication (email) within the US Department of Defense. It wasn't until the mid-1990s, with the introduction of the World Wide Web, that it became widely used.

5.1.2. Henderson and Clark Innovations

Henderson and Clark proposed another framework. Four categories of innovation are distinguished by them.

- **Radical Innovation:** "Establishes a new dominant design and, hence, a new set of core design concepts embodied in components that are linked together in a new architecture;"
- **Incremental Innovation:** "Improves and expands upon an existing design. Individual components can be improved, but the fundamental design ideas and connections between them stay the same;"
- **Architectural Innovation:** "Innovation that changes only the relationships between them the core design concepts; and"
- **Modular Innovation:** "Innovation that changes only the core design concepts of a technology."

Although Christensen, Henderson, and Clark discuss technical innovation, there are other types of innovation as well, including organizational and service innovation.

5.1.3. Other Innovations

- **Non-economic Innovation:** Other forms of innovation include social, religious, sustainable (or green), and responsible innovation, which are different from business-centric perspectives that focus on making money for a company.
- **Open Innovation:** A si□nificant body of literature has focused on open innovation, sometimes known as "crowd sourcing." Using people outside of an organizational setting who are not experts in a particular field to tackle complex challen□es is known as "open innovation."
- **User Innovation:** User innovation is the practice of businesses depending on consumers of their products and services to generate, develop, and even execute new ideas. It is comparable to open innovation.

5.2. INNOVATION AND IPR

Two basic perspectives on innovation have been put forth. In the first, which dates back to Arrow (1962) and was more recently expanded by Romer (1986, 1990), abstract concepts are important. Due to the fact that ideas are non-rivalrous and infinitely replicable at constant marginal cost, which is sometimes taken to be equal to zero, these are generated subject to fixed costs that cannot be recovered under competition. The second, which has its roots in Schumpeter (1941) and was formalized by Boldrin and Levine (2008), holds that the tangible representation of ideas is what has economic worth. Indivisibility is the original characteristic of embodied ideas, but capacity limits limit their replication. Nevertheless, ideas can be produced and sold under competition since the latter generate competitive rents that may offset the indivisibility cost. Both viewpoints contend that "intellectual property" is more than just an extension of "normal" property rights to the domain of ideas; it grants monopolistic power over the ability to reproduce ideas. A monopoly over ideas might be beneficial, according to the first viewpoint, which supports intellectual property not because it is property and protects the value of private investment. Authors like Grossman and Helpman (1991) and Aghion and Howitt (1992) have lately connected this argument to a Schumpeterian theme (Schumpeter, 1942). It suggests a trade-off between "static efficiency," which necessitates competition, and "dynamic efficiency," which can only be attained by technological advancements motivated by the ambition to obtain a monopoly. According to this theory, the creative process is propelled by monopoly power. Three developments have contributed significantly to the controversy surrounding intellectual property legislation.

The first is the exorbitant cost and restrictive practices of pharmaceutical firms, such as those pertaining to AIDS medications. The second is how intellectual property negatively affects the prospects for development of less developed nations, particularly when those nations are excluded from free commerce with more industrialised nations unless they enact stringent intellectual property laws. The effect of the internet on "piracy" of books, films, and music comes in third. On one side of the policy argument are those who profit from current monopolies and are keen to defend their business practices. They usually argue that their "property" should be shielded from "theft." There are a wide range of individuals on the other side who object to authors limiting their copies' freedom of use. In the words of the US Constitution, "To promote the progress of science and useful arts," the main question is whether the monopoly power attained through copyrights and patents is indeed required or if it actually impedes advancement and creativity.

5.2.1. History of IPR

Despite their recent resurgence, intellectual property rights have a fascinating past. The risk of creative ideas being misused and stolen remained present even in ancient times. As a result, ancient thinkers created a policy that closely resembles the current IPR system. The earliest known examples of intellectual concept protection date back to 500 BCE. This policy limited the ability to prepare particular gourmet cuisines to the cooks of the Sybaris' Greek colony. Bruce Bugbee's powerful work from antiquity covered IPR. The incident of Vitruvius was covered in "The Genesis of American Patent and Copyright Law." Poet Vitruvius (257–180 BCE) gained notoriety in Alexandria by reciting the words, poetry, and verses of any other author during a literary competition. Roman poets and painters from the first century C.E. likewise displayed varying ownership interests in relation to IPR. Aside from the protection of these kinds of rights, the ancient Greek or Roman worlds still had any standard written legislation or organisation for the convention of intellectual property protection. However, the use of franchises, privileges, and royal favors prevented the intellectual ideas from being stolen. The main person who gets involved in differentiating between franchises and royal favors is Bugbee. It was noted that royal favors and franchises limit access to ideas that are already in the public domain (Bugbee, 1967). The emergence of the Statute of Monopolies in 1624 and the Statute of Anne in 1710 marked the beginning of the English system of intellectual property protection in America. The Statute of Monopolies granted writers and inventors a 14 year monopoly. But until the invention of the Gutenberg printing press in the 15th century, literary works were mainly unprotected. The first modern copyright statute was established by the Statute of Anne in 1710. By providing 14 year copyrights and a 14 year renewal if the author was still living, the Statute of Anne statute protects authors (Farnley et al., 2004). Belgium, Holland, Italy, and Switzerland were among the European nations that began to adopt the unspoken rule that England had created (Bugbee, 1967). When several international treaties, such as the Berne Convention treaty and the Trade-Related Aspects of Intellectual Property (TRIPS, 1994) treaty, entered the scene, the geographic reach of intellectual property protection expanded around the world.

5.2.2. Various Forms of IPR

Trade secrets, trademarks, copyright, and patents are the four main categories into which intellectual property rights in the 21st century can be readily divided.

5.2.2.1. The Creative Commons, Copyleft, and Licensing

Many online content providers merely offer license agreements (via click-wrap, shrink-wrap, etc.) that permit only particular uses of protected information as a contemporary workaround for the first sale requirement, as opposed to selling a copy of a work. These methods of intellectual property protection are very new and appear to expand on existing copyright frameworks. For instance, the Creative Commons and Copyleft models aim to broaden the commons of thinking and expression by utilising licensing agreements to provide varying degrees of downstream access. If the "new" work is equally functional or accessible, the owner may permit others to build upon a protected work. It should be noted that these agreements and contracts are only morally binding if they are supported by antecedent entitlements, which is implied by the moral arguments presented below. For instance, the resulting agreement or license will not be enforceable if neither party to a Copyleft license owns or has legal ownership to the intellectual work in question. Therefore, ownership or entitlement claims to intellectual works serve as the foundation for the Creative Commons and Copyleft models.

5.2.2.2. Trade Secret

A trade secret is the proprietary information pertaining to market capitalization that allows an entrepreneur to protect the intellectual property rights associated with a specific industrial formula or secret information. Formulas, designs, practices, patterns, commercial methods, tools, marketing, export, and sales plans are all considered trade secrets. Trade secrets are frequently employed in business plans for manufacturing and industry, where the manufacturers who use them maintain the proprietary rights to the product's formula. Trade secrets give their owners financial advantages, typically through commercialization (Patino, 2009). The name of a well-known cold beverage maker, "Coca Cola," is one of the best examples of trade secret keeping. For many years, this company has been able to maintain the recipe as a trade secret without obtaining a patent. If a trade secret is discovered through legal channels, it may be patentable. Compared to patents, trade secrets have less intricate legal circumstances. Although a trade secret offers no defense against abuse, it has an instant impact. Maintaining a trade secret is simple and doesn't involve any legal disclosure or registration fees, but if it is stolen, there is no legal recourse. A trade secret cannot be anything that may be readily revealed by careful product inspection (Rockman, 2004; Becker, 2005). One of the best instances of a trade secret is the Coca-Cola formula.

5.2.2.3. Trade Mark

Another type of intellectual property right that sets a registered provider's product, design, expression, or service apart from rivals or other sources is a trademark. The trademark policy may provide protection for any word, logo, symbol, design, image, colour, scent, sound (such as jingles), or service. The trademark can be bought, sold, or licensed legally. In reality, a trademark is solely used to control the reputation of the commodity or service it represents. Different methods are used to indicate different kinds of trademarks. The symbol for an unregistered trademark that is used to brand or sell goods is ™. The unregistered service mark "TM" is used to advertise services offered by a certain brand. A registered trademark is denoted by the sign ® (Tiwari et al., 2011). During commercial popularization, all trademark and service mark insignia are written as superscripts above the brand names. However, compared to registered trademarks, unregistered trademarks offer less protection against trademark infringement (Figure 5.1).,,

Figure 5.1. *Sample of Logo showing a registered trademark.*

Source: https://www.businesslist.co.ke/company/51140/cargill-kenya-ltd.

5.2.2.4. Copyright

Books, articles, films, short films, reports, diaries, songs, poetry, literature, music, drama, artistic works, paintings, and cinematography are all protected by copyrights, a type of intellectual property rights. There are several advantages to applying copyright through the appropriate way. The circled character © is used to indicate copyright. Copyrights for sound recordings are indicated by the capital P surrounded by the circle . The capital letter R surrounded by a (®) is used to represent a registered trademark, indicating that the product is registered with the trademark office. A capital letter M surrounded by a circle is used as a non-obligatory mark to safeguard the content of the mask art. If the copyright was previously registered with the original creator, the infringement permits the holders of the copyright to be penalized. Software copyright is granted for a finite amount of time. Different countries have different copyright durations. The copyright is valid for at least 30 years and up to 95 years from Yemen to Jamaica. For the duration of the author's life plus 70 years after their passing, copyright protection is a secure method to cover the license associated with the work. The owner of a copyright is granted a number of rights, such as the ability to reproduce, distribute, and enjoy cinematographic rights as well as a portion of the profits made from the work. However, various exclusions were taken into consideration under IPR in order to circumvent the limited dissemination of copyrighted works in the public domain. Extraordinary usage by non-copyright holders is regarded as fair use to safeguard users' interests. A restricted use of the copyrighted content is taken into consideration for religious, educational, research, criticism, and performance purposes (Mann et al., 2010).

5.2.2.5. Patent

The patent is a type of intellectual property right that covers new inventions that are patentable based on their usefulness and novelty while being closely watched and scrutinized. The laws governing patents vary from one country to another, however there are certain commonalities among them. For a patent to be considered, it must be sufficiently unique and beneficial. A patent cannot be granted for an innovation that is deemed frivolous, against the law or morality, or harmful to the public's health. A scientific theory, computation, mathematical formula, or discovery cannot be patented. Patents are not available for inventions pertaining to atomic energy that are effectively customary knowledge. A technique used in agriculture, horticulture, or gardening cannot be copyrighted. In addition to these, there are a number of other items that are prohibited from being patentable, however some of them may be protected by other forms of intellectual property rights (Saha and Bhattacharya, 2011). When a procedure or material is patented, the inventor has a sort of monopoly on the invention, preventing others from using it without permission. Process patents and product patents are two distinct categories of patents. Another way to put it is that the subclasses of patenting systems that are available are utility patents and design patents. The "surface ornamentation," which includes an object's shape or configuration, is referred to as a design. Only the object's appearance is protected by a design patent. Only when the design and the object are inseparable can it be obtained. The type of patent known as a utility patent protects the claimed object's structural or functional aspects. It is the most prevalent and practical kind of patent that may be acquired for new machinery, equipment, production techniques, procedures, and methods. An inventor should apply for both a utility and a design patent if they want to obtain a patent for both the object's functioning aspects and its design (Tulasi and Rao, 2008).

5.2.2.6. Protecting Mere Ideas

A significant body of case law permits people to protect simple concepts as personal property outside of the copyright, patent, trade secret, and trademark regimes. This property system is commonly referred to as the "law of ideas." Buchwald v. Paramount Pictures (13 U.S.P.Q. 2d 1497 (Cal. Super. Ct. 1990)), including the Eddie Murphy film Coming to America, is a well-known case in this field. It was decided that Buchwald would be compensated if a film was produced based on his proposal when he approached Paramount Pictures with it. Copyright infringement did not apply since Buchwald did not correct his notion, for instance, by writing it down. Following years of fruitless talks and failed starts, Paramount informed Buchwald that his proposal would not be made into a film. Eddie Murphy received credit for the release of Coming to America shortly after this notification. Despite the film's purported financial loss, Buchwald filed a lawsuit and won damages. When people come up with ideas and present them to businesses in the hopes of receiving payment, the law of ideas is usually implemented. Compensation may be necessary in some situations when the corporation (or anybody else) uses these concepts without permission. Courts require that an author's notion or ideas be concrete and novel or original before determining that the author possesses property rights to them (Murray v. National Broadcasting, 844 U.S. F2d 988 (Second Cir., 1988)). Only in circumstances of misappropriation is compensation provided (Sellers v. American Broadcasting Co. (11th Cir., 1982)).

5.2.2.7. Droits Morals

The concept of "moral rights," which are included into continental European intellectual property law, is expressed in Article 6bis of the Berne Convention. Known in France as "droits morals" or "moral rights," the idea defends the individual rights of creators, as opposed to their economic rights. These moral rights include the freedom to create and publish a work in any format, the right to claim authorship, the right to stop the work from being deformed, mutilated, or altered in any way, the right to remove and destroy the work, the right to refrain from harsh criticism, and the right to be free from any other harm to one's reputation.

5.2.3. IPR Foundations/Morality

"Everyone has the right to the protection of the moral and material interests resulting from any scientific, literary, or artistic production of which he is the author," states Article 27 of the Universal Declaration of Human Rights. There are ethical justifications for intellectual property, notwithstanding the complicated relationship between it and human rights. Intellectual property, according to personality theorists, is an extension of the individual. Intellectual property, according to utilitarians, encourages social advancement and encourages individuals to continue innovating. Lockeans contend that hard labour and deservingness serve as justifications for intellectual property. Many moral or natural justice-based arguments have been made in favor of expanding IPRs in a variety of fields, including agriculture. Historically, three distinct justifications have been used to support the granting of IPRs which include:

- **Natural Rights/Justice Argument:** Locke's theory that a person has an inherent right over the work and goods that their body produces serves as the foundation for this argument. It is thought to be unfair to appropriate these products. It is conceivable to apply Locke's argument to intellectual property rights, since it would be unfair for anyone to abuse another person's ideas, even if he had never specifically said that natural ri□hts applied to works of the mind. The foundation of Locke's defense of intellectual property is the notion that workers should have authority over the things they produce. They contend that we are the owners of our bodies, which serve as our laborers, and that this ownership extends to the things we produce. Therefore, in terms of production, intellectual property guarantees this right;
- **Utilitarian-Pragmatic Argument:** This argument says that societies that uphold private property rights are more successful and efficient than those that do not. The □rowth of the patent system has been credited for fostering innovation and invention in 19th-century America. Intellectual property laws aim to maximize social usefulness by giving innovators a "durable and tangible return on their investment of time, labour, and other resources." They are

assumed to advance the "creation, production, and distribution of intellectual works" in order to benefit the □eneral public. Accordin□ to utilitarians, there wouldn't be any motivation to generate novel concepts in the absence of intellectual property. Social utility is maximized by protection mechanisms like intellectual property; and

- **"Personality" Argument:** Hegel's statement, "Every man has the right to turn his will upon a thing or make the thing an object of his will, that is to set aside the mere thing and recreate it as his own," serves as the foundation for this argument. The idea that ideas are a "extension of oneself and of one's personality" informs European intellectual property law. According to personality theorists, the act of creating something puts one at risk of having their concepts and designs stolen or changed. These moral assertions pertaining to personality are protected by intellectual property.

Lysander Spooner (1855), for example, contends "that a man has a natural and absolute right—and if a natural and absolute, then necessarily a perpetual, right—of property, in the ideas, of which he is the discoverer or creator; that his right of property, in ideas, is intrinsically the same as, and stands on identically the same grounds with, his right of property in material things; that no distinction, of principle, exists between the two cases" . In her book Capitalism: The Unknown Ideal, author Ayn Rand made the case that intellectual property preservation is fundamentally a moral matter. It is believed that all property is really intellectual property and that the human mind itself is the source of prosperity and survival. Therefore, breaching intellectual property is morally no different from infringing other property rights, which jeopardizes survival processes and is therefore wrong. Countries should be able to create their IPR policies according to their own needs and goals, including those related to agriculture, if an instrumentalist approach is successfully implemented. However, during the past 25 years, a proprietary approach—sometimes linked to moral reward arguments—has impacted both international trends and national le□islation and case law. The field of IPRs applied to plants, animals, microbiological creatures, and their parts and components, including cells and DNA, may contain some of the best illustrations of the impact of proprietarianism.

5.2.4. IPR Infringement, Misappropriation, and Enforcement

Depending on the type of intellectual property involved, the jurisdiction, and the nature of the action, intellectual property rights violations—referred to as "infringement" in the case of patents, copyright, and trademarks, and "misappropriation" in the case of trade secrets—may be considered either civil or criminal violations. In 2011, the global trade in counterfeit copyrighted and branded works was valued at $600 billion and made up 5–7% of all trade. IP has been considered as a means of punishing the aggressor through trade sanctions during the Russian invasion of Ukraine in 2022, has been suggested as a way to stop future nuclear-armed conflicts, and has raised concerns about stifling innovation by keeping patent information secret such as:

- **Patent Infringement:** Using or selling a patented invention without the owner's consent is usually considered patent infringement. The claims of the issued patent specify the degree of protection or the scope of the claimed invention. In many nations, using a patented invention for study is protected by safe harbor. In the US, this safe harbor does not apply unless the research is being conducted for strictly philosophical reasons or to collect information for a drug's application for regulatory clearance. Cases involving patent infringement are often handled under civil law (such as in the US), although certain jurisdictions (such as Argentina, China, France, Japan, Russia, and South Korea) also include criminal law;
- **Copyright Infringement:** Reproducing, distributing, exhibiting, performing, or creating derivative works without the consent of the copyright holder—usually a publisher or other company that represents or is assigned by the work's creator—is known as copyright infringement. It is frequently referred to as "piracy" Although copyright is established in the United States the moment a work is fixed, the owner must often re□ister the copyri□ht in order for the copyright holder to receive monetary damages. The copyright holder is often in charge of enforcing copyright. Signed in May 2011 by the US, Japan, Switzerland, and the EU, the ACTA trade agreement has not yet come into effect but requires its parties to actively police for infringement and impose criminal penalties, such as jail time and fines, for copyri□ht and trademark infrin□ement. Copyright has restrictions and exceptions that permit restricted uses of works protected by copyright that do not amount to infringement. The fair use and fair dealing doctrines are two examples of such concepts;
- **Trademark Infringement:** When one party uses a trademark that is confusingly similar to another party's trademark in connection with goods or services that are similar to or identical to those of the other party, it is considered trademark infringement. A trademark is protected in many nations even if it is not registered, however trademark re□istration offers enforcement benefits. Both civil

litigation and, in some jurisdictions, criminal law can be used to remedy infringement; and

- **Trade Secret Misappropriation:** Since trade secrets are by definition confidential, whereas patents, re□istered copyri□hts, and trademarks are accessible to the general public, trade secret misappropriation differs from other intellectual property law crimes. State laws in the US safeguard trade secrets, and the Uniform Trade Secrets Act has been enacted by almost all states. The Economic Espionage Act of 1996, another federal legislation in the United States, makes it illegal to steal or misappropriate a trade secret. Two provisions in this statute make two different types of activities illegal. The theft of trade secrets for the advantage of foreign countries is illegal under the first section. The second part makes their theft for business or financial □ain ille□al. (The two offences have distinct statutory penalties.) Althou□h trade secrets and confidentiality are considered equitable rights rather than property rights in Commonwealth common law nations, theft penalties are essentially the same as in the US.

5.2.5. IPR International Framework

There are several overlapping organisations and platforms involved in the worldwide governance of intellectual property. There isn't a single body that makes rules. The Agreement on Trade Related Aspects of Intellectual Property Rights (TRIPS) is a crucial component of international IP regulation. Every World Trade Organisation (WTO) member is required to abide by the minimal international standards for intellectual property set forth in the TRIPS Agreement. A member may be sued under the WTO's Dispute Settlement Mechanism for violating the TRIPS Agreement.

5.2.5.1. TRIPS

The TRIPS agreement, which is still the most extensive multilateral agreement on intellectual property to date, was the first to incorporate intellectual property law into the multilateral trading system. The Doha Declaration was the outcome of a series of negotiations that were started in 2001 by developing nations who were worried that rich nations were demanding an unduly restrictive interpretation of TRIPS. A WTO proclamation known as the Doha proclamation explains the parameters of TRIPS, saying, for instance, that it can and ought to be construed in the context of the objective "to promote access to medicines for all." TRIPS specifically mandates that WTO members provide copyright rights, including those of authors and other copyright holders, performance artists, sound recording producers, and broadcasting organisations, as well as holders of related rights, such as geographical indications, industrial designs, integrated circuit layout designs, patents, new plant varieties, trademarks, trade names, and undisclosed or confidential information, such as test data and trade secrets. Additionally, TRIPS outlines dispute resolution, enforcement, and remedy procedures. All intellectual property rights must be protected and enforced in a way that promotes technological innovation, transfers and disseminates technology, benefits both producers and users of technological knowledge, promotes social and economic welfare, and maintains a balance between rights and obligations. IP requirements that surpass those of the TRIPS Agreement are frequently established by bilateral and multilateral agreements. Member governments are required by TRIPS to offer robust intellectual property rights protection. As an illustration, under TRIPS:

- Unless it is based on the author's life, copyright periods must last for at least 50 years;
- According to the Berne Convention, copyright must be conferred automatically rather than through "formality" like registrations;
- Under copyright law, computer programs must be considered "literary works" and be granted the same protections;
- National copyright exceptions, like "fair use" in the US, are limited by the Berne three-step test;
- All "inventions" in all "fields of technolo y" must be iven patents as long as they satisfy all other patentability conditions (though there are some exclusions for specific public reasons), and they must be enforced for at least 20 years;
- As long as there is no contradiction between the regular use of the work and the patent, exceptions to exclusive rights must be kept to a minimum;
- The legitimate interests of the owners of patents and computer programs cannot be unreasonably jeopardized;
- Patent rights must take into consideration the legitimate interests of third parties; and
- According to the concept of national treatment, intellectual property laws in each state cannot provide local residents with any advantages that are not available to residents of other TRIPS signatories (with a few specific exceptions). The most favored nation clause is another feature of TRIPS.

With the exception of moral rights, the TRIPS Agreement references the Berne Convention for the Protection of Literary and Artistic Works' copyright rules. Additionally, it referenced the Paris Convention for the Protection of Industrial Property's substantive provisions. Software and databases are covered by copyright, subject to the originality requirement, according to the TRIPS Agreement.

- The Agreement's Article 10 States:
  - The Berne Convention protects computer programs as literary works, regardless of whether they are in source or object code; and
  - Data or other material compilations that, according to the way their contents are chosen or arranged, qualify as intellectual creations, regardless of whether they are machine readable or not, are protected as such. Any copyright that may exist in the data or material itself will not be impacted by this protection, which does not apply to the data or material itself.

All member states are subject to the same TRIPS responsibilities, but developing nations were given more time to enact the necessary adjustments to their domestic legislation in two stages of transition based on their degree of development. In 2005, the transition phase for developing nations came to an end. The TRIPS transition period for least developed nations was extended to 2013, and for pharmaceutical patents, it was extended to January 1, 2016, with the potential for an additional extension. Therefore, it has been argued that the TRIPS standard, which requires all nations to establish stringent intellectual property laws, will hinder the development of less developed nations. It has been ar□ued that most, if not all, developin□countries would benefit strate□ically from usin□ TRIPS' flexibility to enact the weakest IP rules possible. In the majority of cases, this has not occurred. According to a WHO report from 2005, many developin□nations have not fully implemented the TRIPS flexibilities—such as parallel importation, mandatory licensing, data protection restrictions, the use of broad research, and other patentability exceptions—into their laws as permitted by Doha. This is probably because developing nations frequently copy the IP laws of developed nations directly or rely on technical assistance from the World Intellectual Property Organisation (WIPO), which critics like Cory Doctorow claim encourages them to impose stronger intellectual property monopolies. This is likely due to a lack of legal and technical expertise required to draft le□islation that implements flexibilities. TRIPS has a favourable impact on Indian pharmaceutical companies' R&D expenditures, as demonstrated by Banerjee and Nayak.

Academics, non-governmental organisations, and developing nations have all criticized TRIPS since it went into effect. Many proponents of trade liberalization also view TRIPS as bad policy, even though part of this criticism is directed at the WTO in general. Such criticisms are frequently based on TRIPS' wealth concentration effects, which transfer funds from individuals in developing nations to copyright and patent holders in developed nations, as well as its imposition of artificial scarcity on the citizens of nations with weaker intellectual property laws. The inability of TRIPS to speed up technolo□y and investment flows to low-income nations—a benefit promoted by WTO members prior to the agreement's creation—has been the subject of additional criticism. According to World Bank statements, TRIPS may have accelerated investment in middle-income countries, but it hasn't shown any signs of accelerating investment in low-income nations. According to Daniele Archibugi and Andrea Filippetti (2010), TRIPS was primarily implemented to counteract the deterioration in the competitiveness of the US, EU, and Japanese technology industries against emerging countries. Instead, they contend that IP-intensive multinational firms in these nations were the primary proponents and recipients of TRIPS, as it allowed them to outsource important functions to developing nations.

Additionally, Archibugi and Filippetti contend that proponents of TRIPS have exa□□erated the si□nificance of intellectual property in □eneral and TRIPS in particular in the creation and spread of knowledge and innovation. The United Nations has shown that many nations with inadequate protection often reap the benefits of hi□h levels of forei□n direct investment (FDI), which lends credence to this notion. An examination of OECD nations in the 1980s and 1990s, when pharmaceutical patents were extended by six years, revealed that although the overall number of items registered somewhat increased, the mean innovation index stayed constant. On the other hand, Jörg Baten, Nicola Bianchi, and Petra Moser (2017) discover historical evidence that, in some situations, compulsory licensing—a crucial tool for undermining intellectual property rights covered by Article 31 of the TRIPS agreement—may actually be useful in encouraging invention by raising the threat of competition in industries with low levels of competition already in place.

However, they contend that the advantages of lowering intellectual property rights mostly depend on whether governments can legitimately promise to use it only in dire circumstances, as businesses may be less inclined to engage in R&D if they anticipate frequent instances of compulsory licensing. There has also been examination of TRIPS-plus provisions that require requirements above and beyond TRIPS. Governments' capacity to create competition for generic producers is restricted by provisions in these free trade agreements. The United States in particular has come under fire for promotin□protection that goes much beyond what TRIPS requires. Patentability has been expanded by the US Free Trade Agreements with Australia, Morocco, and Bahrain, which mandate that patents be granted for novel applications of well-known goods. At a country's discretion, compulsory licenses may be granted under the TRIPS agreement. The United States' free trade agreements with Australia, Jordan, Singapore, and Vietnam have TRIPS-plus clauses that limit the use of compulsory licensing to public non-commercial usage, emergency situations, and antitrust remedies.

5.2.5.2. WIPO

The WIPO is a global organisation that seeks to advance the protection of copyrighted items (such as books, music, photographs, and other artistic creations) and industrial property (such as inventions, trademarks, and designs). Established by a 1967 Stockholm convention, the organisation started activities in 1970 and was officially recognized as a United Nations specialized agency in December 1974. Geneva is home to its headquarters. The Paris Convention for the Protection of Industrial Property, which established intellectual property protections for innovations, trademarks, and industrial designs, was signed by 14 nations in 1883, marking the beginning of WIPO. Outside of their own nations, inventors were able to protect their creations thanks to the convention. The Berne Convention of 1886 mandated that member nations automatically safeguard works created in other member nations. After establishing independent secretariats to implement their own treaties, the two organisations united in 1893 to form the United International Bureau for the Protection of Intellectual Property (BIRPI), with its headquarters located in Bern, Switzerland. BIRPI relocated its headquarters to Geneva in 1960. WIPO has two objectives. First, WIPO encourages intellectual property protection through global collaboration. Over 20 intellectual property treaties are now administered by the organisation. Second, with relation to trademark, patent, and artistic and literary work protection agreements, WIPO oversees administrative cooperation between the Paris, Berne, and other intellectual unions. When WIPO and the World Trade Organisation signed a cooperation agreement in the mid-1990s, WIPO's role in enforcing intellectual property protections expanded. WIPO was tasked with assisting in the settlement of disputes pertaining to the use of Internet domain names as electronic commerce expanded with the growth of the Internet. More than 180 countries are members of WIPO. Every two years, the General Assembly meets as its principal policy-making body. The organization's budget and programs are decided during WIPO's biennial conference. Observer status is maintained by about 170 nongovernmental organisations.

In the Council for TRIPS on the revision of Art. 27.3(b), the first substantial discussion of elements of the international intellectual property law that affected food security took place. In the context of agitation for the inclusion of traditional knowledge into the international intellectual property framework, this topic went to the WIPO when the Seattle Ministerial failed in November 1999. As part of the discussion on "Matters Concerning Intellectual Property and Genetic Resources, Traditional Knowledge and Folklore" in the WIPO General Assembly, the Dominican Republic's Permanent Mission to the UN in Geneva submitted two documents on behalf of the Group of Countries of Latin America and the Caribbean (GRULAC) in a note dated September 14, 2000. The establishment of a Standing Committee on Access to Genetic Resources and Traditional Knowledge of Local and Indigenous Communities was the main emphasis of these publications. "That Standing Committee's work would have to be focused on establishin□ □lobally accepted workable strate□ies for □uaranteein□ sufficient protection for the intellectual property rights in traditional knowledge." It was proposed that the Committee provide clarification on the followin□ points to help understand how intellectual property will be applied in the futurc to the use and exploitation of biodiversity, genetic resources, and traditional knowledge:

- The notions of public domain and private domain;
- The viability and propriety of granting rights to classical works and information that are currently in the public domain, as well as lookin□ into mechanisms to restrict and mana□e specific types of unapproved utilization;
- Recognition of collective rights;
- The principles of prior informed consent and equitable profit distribution in relation to the use, development, and commercial exploitation of the transferred material and inventions are embodied in model provisions and model contracts that regulate the use and exploitation of genctic and biological resources, as well as machinery for the equitable distribution of profits in thc event that a patentable product or process is developed from a given resource. and technology resulting from it; and
- The protection of undisclosed traditional knowledge.

5.2.5.3. Intergovernmental Committee

The creation of an Intergovernmental Committee on Intellectual Property and Genetic Resources, Traditional Knowledge, and Folklore was approved by Member States at the WIPO General Assembly. Three interconnected topics were found to guide the Committee's discussions: problems with intellectual property that come up in the context of:

- Access to □enetic resources and benefit sharin□;
- Preservation of customary wisdom, whether or not it is connected to those resources; and

- The protection of expressions of folklore.

The Intergovernmental Committee on Intellectual Property and Genetic Resources, Traditional Knowledge and Folklore ("IGC") held its First Session in Geneva from April 30 to May 3, 2001. During this session, the Member States prioritized certain tasks for the Committee and decided on the agenda items on which work should be done. "The development of 'guide contractual practices,' guidelines, and model intellectual property clauses for contractual a□reements on access to □enetic resources and benefit-sharin□' was the main one of these. The resources listed below were created by the Fifth Session of the IGC, which took place in Geneva from July 5–15, 2003, and were pertinent to the topic of food security:

- An examination of case studies on the use of IP to protect traditional knowledge carried out by WIPO between 1998 and 1999, as well as a comprehensive study of the protection of traditional knowledge using IP laws;
- Concerns around IP management that come up while recording traditional knowled□e are identified in A Draft Toolkit on Intellectual Property Management 79;
- A list of terms and procedures for contracts that deal with intellectual property, benefit-sharin□, and access to □enetic resources;
- A Technical Analysis of Traditional Knowledge and Genetic Resource Disclosure Requirements. In addition to reviewing key elements of the legal framework and patent system pertaining to access to genetic resources and related traditional knowledge, this study also examined the answers to a survey on patent disclosure requirements that was distributed to WIPO Member States; and
- Based on the findin□s of the WIPO Asia Pacific Re□ional Seminar on Intellectual Property and Genetic Resources, Traditional Knowledge and Folklore, which took place in Cochin, India, from November 11–13, 2002, the Asian group of countries submitted technical proposals on databases and registries of traditional knowledge and biological/genetic resources.

From March 15–19, 2004, the WIPO Intergovernmental Committee on Intellectual Property and Genetic Resources, Traditional Knowledge, and Folklore convened in Geneva. Regarding genetic resources, it took into account:

- **Draft Intellectual Property recommendations for Access and Benefit Sharing Agreements:** The recommendations were designed to help in the development of a□reements for benefit-sharin□ and access to genetic resources and related data, including traditional knowledge.

- **Requirements for Genetic Resources and Patent Disclosure:** A text on procedures for handling disclosure of information about genetic resources in patent applications was suggested by the Committee's Draft Technical Study on Patent Disclosure Requirements Relevant to Genetic Resources and Traditional Knowledge, which was sent for adoption by COP at its seventh meeting (9–20 February 2004):
- **Defensive Protection Strategies for Traditional Knowledge, Genetic Resources, and Intellectual Property:** The following topics were examined under this topic: how to better document genetic resources and traditional knowledge for use in patent procedures; how to better understand innovations within traditional knowledge systems for the purposes of patent search and examination; and whether or not oral disclosures of traditional knowledge should be recognized. In its further work, the Committee suggested taking into account:
  - "The compilation of information regarding the criteria that apply to the determination of relevant prior art in various jurisdictions so that the intended objectives would be achieved when defensive publication is made for patent purposes"; and
  - The creation of "recommendations or guidelines for national patent offices concerning searches in the area of inventions linked to TK (within specific technical fields) or genetic resources," with the intention of making sure that patent authorities who lack experience with traditional knowledge systems are better able to base their decisions on a better understanding of how TK is developed and maintained in the traditional context.

5.2.5.4. Substantive Patent Law Treaty

A draft Patent Law Treaty (PLT), created by the International Bureau of WIPO, was examined by a Committee of Experts and WIPO's Standing Committee on Patents (SCP) in an effort to come to an agreement on important patent law matters. The Draft PLT addressed a number of patenting-related procedural issues. In order to achieve some worldwide harmonization of patent registration procedures, the Colombian delegation requested the inclusion of the following chapter in the PLT during the third session of the SCP, which took place from September 6–14, 1999:

- The preservation of the nation's biological and genetic legacy must be ensured by all industrial protection. As a result, the granting of patents or registrations pertaining to aspects of that legacy will be contingent upon their lawful acquisition; and
- The registration number of the contract that grants access to genetic resources and a copy of the contract that states that the products or processes for which protection is sought were developed or manufactured from genetic resources, or products thereof, of which one of the member countries is the country of origin, must be specified in every document.

Whether this proposition originally presented a procedural or substantive patent law issue was the subject of a contentious discussion. In the end, it was decided to postpone this proposal's consideration until after the proposed Substantive Patent Law Treaty was discussed. A Working Group on Biotechnological Inventions was to be held at WIPO in November 1999, and the SCP asked the International Bureau to put the protection of biological and genetic resources on its agenda. At its meeting the following month, the Working Group suggested the creation of nine initiatives pertaining to the safeguarding of biotechnology inventions. The Working Group made the decision to create a questionnaire in order to collect data regarding the protection of biotechnological inventions in WIPO Member States, includin□ specific aspects pertainin□ to intellectual property and genetic resources.

5.3. IPR IN APPLICATION

5.3.1. IPR in Biodiversity and Bio-Resources

The varied forms of many creatures, including their genetic diversity worldwide, are referred to as biological diversity. For human society, this natural biological wealth and the interactions between genes, species, and ecosystems are invaluable. We owe the glittering bio-resource of millions of plants, animals, fungus, and other species. Populations of organisms, genetic resources, and other biotic elements of ecosystems are examples of biological resources that have substantial potential utility or worth for humankind. The question of whether or not biodiversity and bioresources should fall under the purview of IPR was up for dispute. The current discussions surrounding the incorporation of these properties under the purview of intellectual property rights had previously left the Traditional Knowledge of Biodiversity unaffected. As the demand for bioproducts increased annually, so did the worries regarding the incorporation of indigenous resources, genetic resources, and bio-resources. Bioresource coverage under IPR is in fact a global necessity. Traditional knowledge relating to biodiversity has been commercialized globally in recent decades thanks to international accords like the Convention on Biological Diversity (CBD) and the TRIPS. Countless animal species, genetic resources, plants, trees, and microbes that have potential uses of commercial and economic significance are found in biorich nations like India and the Third World. Covering the use of bioresources under IPR becomes crucial in order to prevent the major danger to biodiversity and bioprospecting. Of the 1.75 million known species in the global biodiversity pool, 2.7 lakh are found in the kingdom of plants. It is difficult to manage such a vast pool of plant and animal bioresources with equitable distribution and to avoid exploitation. Animal and plant species are distributed unevenly and in a variety of ways throughout the world. Equitable resource distribution between nations with high biodiversity and regions with moderate and low biodiversity is extremely challenging. These organisms, which are crucial for trade, medicine, food, fashion, tourism, energy, and leisure, are now protected by intellectual property rights. The common heritage of humanity, biological diversity, patentable microorganisms, plant variety protection, farmers' rights, and national sovereignty—including conservation, sustainable use, fair benefit sharing, etc.—are all covered by a number of acts and laws that fall under the purview of intellectual property rights in the bioresource sector.

5.3.2. IPR in the Pharmaceutical Industry

Compared to other biotechnology and medical science-based commercial benefits, the pharmaceutical industry is expanding quickly. Maintaining the drug's quality while launching new medications for newly emerging illnesses is extremely competitive. These days, the pharmaceutical business is more interested in long-term consumable medications than in short-term ones. While there is a sizable market full of piracy, it is extremely expensive for the corporation to bring the new medication to market. Obtaining a patent or intellectual property covering for any novel or generic formula is crucial in this rapidly globalized market. In the past 10 years, the number of applications to protect IPRs pertaining to the pharmaceutical and drug industries has nearly tripled. As a result, the appropriate authorities are quite selective when it comes to awarding patents and providing IPR protection for novel medications. These are the causes of the ensuing decline in the number of years that patent protection is offered. The pharmaceutical sector must also balance a number of competing needs, such as maintaining cheap prices to satisfy consumer desires. This is a challenging undertaking because to biotechnology research and development, production, management, and marketing chains. Due to intense global rivalry, the pharmaceutical industry typically waits to publish an idea or formula until it has been properly patented. By protecting the drug formula from theft or copying, patenting allows the inventors and true owners to keep their economic rights. Compared to straightforward traditional medications, the process of intellectual property protection is somewhat more complex for biotechnologically created biopharmaceuticals. Prior to IP protection, every microbial strain used in development should be identified; if the strain is already known, the database should be taken into consideration; if it is new, it should be deposited and registered in the relevant repository before the medicine is patented. Since protecting IPRs in the pharmaceutical sector is a complex undertaking, there are numerous restrictions and challenges with pharmaceutical industry patents. It takes specialized professional, scientific, legal, and technological knowledge to write patents in the pharmaceutical sector. New combinations and inventions that have never been used before could potentially be patented by the inventor. It may also be possible to patent the combination of pre-existing materials to create something beneficial.

5.3.3. IPR in Bioinformatics and Computational Biology

The field of study known as bioinformatics uses computers or computer-based information technology to develop a clear, multidisciplinary method for managing biological data and information. The necessity for bioinformatics in biological science research has grown dramatically since the structure of DNA and RNA was discovered. The largest result of bioinformatics in the present world is the creation and administration of diverse software-based scientific data. The creation, multiplication, and management of molecular science data—which includes gene, DNA, RNA, protein, and functional sequence-based data—make the origins of bioinformatics increasingly important every day. It takes a lot of time, money, and effort to understand biological techniques, develop new pharmaceuticals, and detect and identify different gene-based therapies; for this reason, it is crucial to cover both new and existing findings inside a legal framework. IPR protection for bioinformatics results guarantees that the true creators will continue to reap the benefits and financial gains for the rest of their lives. One of the quick fixes to guarantee IPR protection in the bioinformatics field is the patenting system. In this manner, the patent holder maintains the right to promote and commercialize the product without worrying about theft, copying, or misuse. Copyright protection and trademark development also address the security of fundamental rights pertaining to economic reasoning. Personalised studies based on genes and the genome can have limits, though. Depending on how it is presented and what information it contains, any database may or may not be a patentable topic. On the other hand, software is more likely to be patentable for the developers, allowing them to profit from its commercialization. Manufacturers can utilise IPRs to protect the practical, physical, and tangible information that a certain piece of software offers. With the advent of numerous computer tools for research, synthetic biology has expanded significantly in recent years. Under these kinds of studies, life-constructing pathways are examined at the genomic level. These initiatives surpass the scope and magnitude of conventional recombinant DNA technology and related methodologies. One unconventional field that has serious problems with IPR protection is biological engineering. Computers and computational procedures face copyright and patent issues, while biotechnology-related approaches have the most challenges in terms of patent law. Original works of expression are protected by copyright, which specifically excludes functional works. However, formulas and algorithms are typically excluded under patent law, which takes functionality into account. As a result, many computerised methods do not satisfy the requirements of the patent box or copyright. It was too useful for copyright and too similar to a set of concepts and algorithms for a patent. A few well-known foundations Monoclonal antibodies and recombinant procedures are examples of biotechnological results that were either unpatented or patented. If patented, the methods and advantages were often made freely accessible at a fair price to the patent holders. One thing is certain, though: between software and hardware inventions, the law and IPRs continue to struggle.

5.3.4. IPR in Biotechnology and Corporate Biotechnology Transfer

With vast growth and development potential, biotechnology is a sister field of biology. Humans have been employing biotechnology in a quiet way since the beginning of time, but in the last 10 years, the introduction of genetic alteration in plants, animals, and microbes, as well as the development of unconventional techniques, have allowed this branch to blossom into a beautiful flower. Due to the increasing number of new ideas, it is imperative that innovators present their creations with intellectual property rights. The importance of IPRs in the corporate world of biotechnology is supported by a number of factors. Depending on the topic, all of the common IPRs—patent, copyright, trade secret, and trademark—apply equally in the biotechnology sector. IPRs in the biotechnology sector in India are likewise subject to a number of restrictions. Any plant species cannot be patented in India. In a similar vein, software developed in India cannot be patented; nevertheless, it can still be protected by copyright. The legal advantages of plant species that cannot be patented are covered under the protection of plant breeders' rights. With their current alternative support systems, the corporate biotechnology sector has numerous other constraints that can be overcome. Because the biotechnology sector is limited in its ability to produce new innovations on a daily basis, it needs an updated and pertinent IPRs framework. IPRs greatly aid in generating income in addition to safeguarding materials, inventions, processes, information, and technology from unauthorized usage. When a license for patented technology or material is granted, the owner receives royalty in two ways. By granting authorization for fair use, the owner can profit from license sharing and royalties; but, if the usage is unfair, the illegal users must pay compensation. The largest source of funding for organisations that employ scientists, researchers, and inventors is revenue. This steady stream of money keeps coming in. The revenue from patented or copyrighted products will cover the significant time, money, and effort required for the research and development branch. This procedure keeps the steady money available for the industry's continued survival.

In addition, IPRs support the inventors' moral and psychological reward system in addition to provide financial and legal protection. If the innovators are not recognized, renowned, or compensated for their prior labors, it is truly worthless. IPR patenting and the copyright system allow individuals and organisations to keep their brand, reputation, and income. Because they have a large enough platform to demonstrate their abilities for both their personal and society's benefit, this makes the inventors feel more relaxed and comfortable mentally. In biotechnology, nucleotide sequences can be patented for a few years before becoming public domain; gene patenting schemes are also in place in a few nations; however they are controversial. However, the non-patentability method often covers the genes and cells of humans and animals. However, there are several exceptions. Genetically modified microorganisms are patentable. In 1985, the United States patented genetically modified crops, including "tryptophan overproducing maize." Likewise, the "on-como-use" was patented in the United States in 1988 due to its genetic modification. Another kind of nonconventional patent that differs slightly from a traditional utility patent is a plant patent. A plant that is produced through asexual reproduction is eligible for a plant patent, but the tuber-propagated plant is not. In biotechnology, which is widely used for hybridisation methods, cell line processing, customer lists, and corporate merchandising strategies, trade secrets are another common IPR type. The Berne Convention on copyright was reflected in a 1999 update to the Indian Copyright Act, 1957. The Berne Convention, Geneva Convention, WIPO, Geneva, and UNESCO are just a few of the agreements and organisations that India is a party to.

5.3.5. Biological Database and IPR

Libraries-like information sources about current biological data are called biological databases. Based on the information, the biological database could be categorised into a number of sub-divisions, such as sequence, structural, and functional databases. Important responsibilities in IPR management are also played by all primary databases (GenBank, EMBL, DDBJ, SWISS PROT, PIR, PDB), secondary databases (UniGene, PRINTS, BLOCKS), and composite databases (NCBI). While secondary databases typically entail capital, labour, time, etc., and are therefore regarded as being protected under IP protection, primary databases are typically gathered by government or social financing resources; as a result, they are typically open to the public without any IPR protection.

5.4. GENERAL CRITIQUES OF IPR

There are a number of broad criticisms of the right to regulate intellectual property to take into account, putting aside the lines of reasoning that attempt to support moral claims to intangible works and the fairly specific issues with these viewpoints.

5.4.1. Information is Not Property

Critics contend that, according to the conventional definition, information is not the type of thing that can be owned, possessed, or considered property. Since they are abstract and unable to causally interact with material objects, information objects—like numbers and propositions—cannot be owned or possessed. For instance, the notion that one might possess and therefore own the novel represented by the book A Tale of Two Cities is as illogical as the notion that one could possess and thus own the entity represented by the number "2." According to these thinkers, the concept of property does not apply to abstract things in the same way that other conceptions could. The word "intellectual property" is conceptually meaningless at best and illogical at worst. There are at least two problems with this analysis. First, it's unclear if ownership as a notion necessitates actual possession. One could contend that ownership is more about having the authority to bar others from engaging in specific activities affecting the relevant entity than it is about having actual control or possession of the thing. Second, the argument that information objects cannot be property does not mean that giving writers or content creators the legal authority to prevent others from using those things without their permission is illegal. It does not follow that "E" should not be protected in very similar ways; rather, the fact that some entity E is not "property" simply means that it should not be legally protected as such. Perhaps these legal rights should be referred to as anything other than "intellectual property rights," although they could also be known as "intellectual content rights."

5.4.2. Information is Non-Rivalrous

Many have claimed that there is a prima facie case against the right to restrict access because intellectual works are not rivalrous. We have a solid argument against moral and legal intellectual property rights because intellectual works are not usually consumed by their usage and can be used by multiple people at the same time (creating a copy does not deprive anyone of their belongings). Many people believe that limiting access to intellectual works is unfair, which contributes to the widespread piracy of these works. The first premise is the weakest part of this argument (Moore 2012; Himma, 2005b). Think about private, sensitive data. Moore contends that it is incorrect to assert that a prima facie moral claim to maximal access is established just because this knowledge can be accessed and consumed by numerous people at the same time. Obscene pornography, snuff videos, national security information, private financial information, and private ideas are all non-rivalrous, but this does not automatically create prima facie moral claims for maximum access and use. Furthermore, even in situations where the copier would not have lawfully acquired a copy (and so is not depriving the owner of the financial recompense they would otherwise receive), it is unclear that unauthorized copying does not cause injury to the owner. Unauthorized copying exposes owners to hazards they did not consent to. Himma notes that the assertion that information consumption is non-rivalrous does not, by itself, suggest that we have any sort of claim to those items. Although this is undoubtedly a reason not to believe that intellectual property protection is morally justified, it does not tell us whether we have any kind of right because it contains no information about morally significant human characteristics. Furthermore, the justification of general rights claims must rely on value attributions that implicitly address the interests of beings with the proper level of moral standing, in our case, our status as persons (Himma 2005b).

5.4.3. Information Wants To Be Free

According to Barlow (1997), information has a moral claim just by virtue of its existence. Information, in his opinion, is a sort of life that claims to be free and is based on its own interests and "wants." Information objects, as he states, "are life forms in every respect but a basis in the carbon atom." They reproduce themselves, engage with their environment and adjust to it, change, and endure. Furthermore, there is some interest in making these live information items freely accessible to everyone. There are a few ways to refute Barlow's reasoning. First, Himma (2005b) contends that the idea that abstract objects have desires or even interests is just not realistic. The idea of want is such that only conscious beings can have wishes; non-sentient entities are no more appropriately described as having desires than as having hopes, despite the fact that conscious beings can have latent desires. Second, Barlow provides no justification for assuming that information objects have a want or interest in being made publicly available, even if they do have needs or interests. Undoubtedly, the assertion that information objects are somehow benefited by being made publicly available requires support, if only because it defies logic.

5.4.4. The Free Speech Argument Against Intellectual Property

Some argue that upholding intellectual property rights conflicts with our dedication to free speech and thought. The idea that people have a right to information and that intellectual property organisations infringe upon this fundamental right is closely related to this argument. Like trade secrets, Hettinger contends that intellectual property "restricts methods of acquiring ideas, restricts the use of ideas, and restricts the expression of ideas, like copyrights—restrictions undesirable for a number of reasons" (Hettinger, 1989). Hettinger identifies trade secrets as the most problematic since they are exempt from disclosure requirements, in contrast to patents and copyrights. This type of concern has been addressed with three different types of responses (Himma, 2005; Moore, 2012). Although we concentrate on the free speech argument against intellectual property, arguments for the right to know are also subject to comparable criticisms. According to the first objection, a system of intellectual property protection may temporarily restrict access, but overall it improves the freedom of thought and expression. This is because the incentives provided by limited protection encourage the production and sharing of information. In other words, the incentives provided by copyrights and patents ultimately give us more to discuss, consume, and enjoy. Second, it is not entirely certain that free speech is so presumed to be important that it almost always takes precedence over other values. We would not tolerate shouting at someone all day long over a bullhorn as protected free speech. Examples of speech that we are ready to restrict for a variety of reasons include hate speech, pornographic expressions, sexual harassment, and the dissemination of private medical information about others. Perhaps intellectual property rights can be seen in this context. Lastly, think about the controversial but well-established copyright idea/expression rule. Ideas that might constitute a fixed expression are not protected by copyright; only fixed phrases are. For instance, someone could study Darwin's original works on evolution, reword them, and then get a copyright for the new version. Even though this person may have plagiarized, she is still able to get a copyright as long as her words are not taken directly from Darwin's original work or significantly comparable to it. Since patents and copyrights demand disclosure, the concepts that comprise these works of intellectual property can be examined and debated.

5.4.5. Social Nature of Information Argument

This perspective holds that information is a communal good and that authors and inventors gain undue advantages from access restrictions. People grow up in civilizations that impart information, which they subsequently utilise to produce a wide range of intellectual works. According to this perspective, knowledge—the foundation of intellectual works—is a social product. Since their creations are based on the collective knowledge of society, individuals shouldn't have sole and permanent ownership of them. It would be like giving ownership to the person who laid the final brick in a public works dam if intellectual property rights were granted. Just as the dam was constructed by hundreds of people, knowledge, the foundation of all intellectual endeavour, is also constructed in a similar way. Critics have questioned not just whether the concept of "society" used in this approach is explicit enough to carry the weight that the reasoning demands, but also whether societies may own or deserve something, or be owed something (Spooner, 1855; Moore, 2012). According to Spooner, "it would probably be very difficult to define what rights society has in ideas that they did not produce and have never purchased; and equally difficult to explain how society came to possess those rights." To demonstrate that society has any legitimate claim to specific concepts—the results of individual labor—or, hence, any rights in them, it is undoubtedly necessary to do more than merely declare them (Spooner, 1855). The issue of how societies gain property rights is now the same as the issue of how individuals acquire them (personality, utilitarian, Lockean, etc.). The proponent of the "shared culture" view is guilty of a similar trick to the defender of the first cause argument for the existence of God, who rides the principle of sufficient causation to a certain point and then conveniently abandons it (every event or object needs a sufficient cause and nothing is self-caused except God). Moore accuses those who support the social nature argument against intellectual property of failing to see that it may prove too much. According to Moore (2012), "shared culture" or the social nature of intellectual property is adequate to undermine intellectual property rights or strong control over creative works, but it is opportunistically insufficient to undermine student desert for a grade, criminal punishment, or other moral evaluation. Lastly, it may be claimed that we have already paid for the use of this common expertise when we pay for schooling and similar services, even if a supporter of this viewpoint can defend society's ownership of general pools of knowledge and information.

5.4.6. Cost of Publishing Digital Information

According to Coy (2007), the price of information should accurately reflect the cost of providing it to consumers in a market that is competitive. According to this line of reasoning, the cost of publishing information in traditional material media, such as books, may be high enough to warrant charging users a fee, but as the number of users increases, the cost (per user) of making information available on digital media approaches zero. For instance, posting information on a website may have some fixed costs, but making that content accessible to any number of users doesn't require any extra money; the more users that appropriate the information, the less expensive it is to make it available to any specific user. Therefore, the argument concludes that charging users for appropriating any piece of (digital) information would be unfair; instead, information should be free (or almost free) to represent the costs of transmission. This argument is flawed in two ways. First, the voluntary interactions of buyers and sellers in a competitive market will determine what a fair price is if one accepts the legitimacy of free enterprise, as the aforementioned argument seems to imply. The fair price is the one that is established by the contractual transactions of free, prudentially-rational buyers and sellers. It can be assumed that a price is reasonable if consumers in a competitive market are prepared to pay a price for digital information that is far greater than the seller's marginal cost. Second, the reasoning ignores the fact that creating and disseminating intellectual property might have significant fixed costs. For instance, the Disney Company invested almost $100 million to produce the 2021 movie West Side Story. It is reasonable for content creators to charge a price that is high enough above the marginal costs to enable them to recoup the fixed development costs involved in creating and disseminating intellectual property, if one assumes that a fair price is one that does so.

5.4.7. Liberty Objection to Intellectual Property

When someone possesses a tangible object, her rights prevent others from interfering with her authority over it. However, intellectual property rights limit personal freedom even in the privacy of one's own home, encompassing all spheres of human activity. "How can the owner of the patent, copyright, or artist decide what I can do with my stuff?" The idea of intellectual property has been criticized by many for violating people's rights to liberty (Palmer, 1990). According to Tom Palmer, intellectual property and liberty appear to be at odds because, although property in tangible objects restricts actions only with regard to specific goods, property in ideal objects restricts a wide range of actions that are not limited by time or place and involve legally owned property (such as VCRs, tape recorders, typewriters, the human voice, and more) by everyone except those who are privileged to receive monopoly grants from the state (Palmer, 1990). This kind of concern has at least two responses. First, the issue could be resolved by enacting particular laws. For instance, clauses that restrict the rights of authors and inventors are ingrained in the contemporary Anglo-American intellectual property structures. Certain restrictions, such as "fair use" (17 U.S.C. §107) and "first sale" (17 U.S.C. §109(a)), permit people to use a patented or copyrighted work for non-profit, educational, or personal use. Replicating a chapter of a library book or creating backup copies of computer games are both legal under current law. These kinds of restrictions might be incorporated into the agreement between society and intellectuals. A second response is that people's freedoms to use their bodies and possessions are limited by various rights. Intellectual property does not restrict individual liberty any more than other types of rights do. For instance, a car's physical property rights forbid someone from getting angry and causing harm to the vehicle. It is illegal to hit someone in the face, damage a guest's personal belongings, or take part in dangerous actions that endanger one's neighbors, even in the quiet of one's own house.

5.4.7.1. Note

IPRs are incredibly helpful, but only for people who are aware of their rights and act to protect them. Determining early on whether any knowledge or concept is necessary to safeguard under the legal line of defense employing IPRs is highly advantageous. As soon as a professional action is taken to protect the concept, service, or product, the rights are better protected from infringement or competition. The kinds of IPRs that are appropriate for a certain situation should be carefully determined. A fundamental requirement for improved management and commercialization of biotechnological applications is an understanding of the IPR. Globalizing the commercialization of technology, creativity, or invention without worrying about theft or misuse is made possible by the identification of rights and careful planning to protect intellectual property. IPR is essential to the administration of creativity and invention protection. It is imperative that IPR policies be updated and incorporated as a foundational strategy into management systems. A prudent choice made early on safeguards a secure and prosperous business future.

Chapter 6

IPR in Food and Agriculture

Contents

OVERVIEW

Legally granted rights pertaining to specific kinds of economic, scientific, and artistic creation are known as intellectual property rights, or IPRs. By creating a defined period of commercial exclusivity during which research and development (R&D) costs can be recovered, the main policy goals for granting IPRs are to incentivize investment in innovation and to establish a legal framework that promotes technology transfer. IPRs like patents, plant variety rights, trademarks, and geographical indications are at the core of the majority of technology transfer agreements. These are usually bundled with related know-how agreements, raw material supplies, and technical and commercial services like marketing and advertising. IPRs are essential to contemporary agriculture. By producing tradeable assets and acting as security for investments in those assets, they help to make agricultural R&D more appealing. The following are the main IPRs that are pertinent to agricultural innovation:

- Patents, which protect inventions;
- Plant variety rights, which protect the breeding of new and distinct plant varieties; and
- Trademarks and geographical indications, which facilitate the marketing of products by providing protection for the symbols of their manufacturing or geographic origin.

Also relevant, but of lesser significance, are;

- Industrial designs, which protect the aesthetic appearance of products, such as agricultural machinery;
- Layout designs of integrated circuits, which are relevant to smart agricultural technologies;
- Confidential information law, which protects know-how or trade secrets; and
- Copyright, which protects works of cultural creativity, such as books, articles, scientific papers, and arrangements of data.

The Papal States Edict of September 3, 1833, on the declarations of ownership of new inventions and discoveries in the sectors of agriculture and the industrial arts, was the first legal proposal for the protection of agricultural innovations (Laclavière, 1962). The Paris Convention for the Protection of Industrial Property, which was adopted in 1883, defined "industrial property" to include agriculture. According to Article 1(3) of the Convention, industrial property applies to all manufactured or natural products, such as wines, grain, tobacco leaves, fruit, cattle, minerals, mineral waters, beer, flowers, and flour, and it encompasses not only industry and commerce proper but also agricultural and extractive industries. Given the state of technology in 1883, the protection of trademarks and source indications was most likely the reason these agricultural topics were included in the definition of "industrial property." The United States Congress introduced a "Bill to amend the laws of patents in the interest of the ori□inators of horticultural products" in 1906, which was the first national proposal that hinted at the protection of agricultural discoveries under patent law. Similar measures were introduced in 1907, 1908, and 1910, but none of them succeeded. The "Plant Patent Act," also known as the Townsend-Parnell Act of 1930, was the first piece of le□islation that acknowled□ed a□ricultural discoveries. In an effort to bring their patent systems into line with the United States,' Cuba (1937), South Africa (1952), and the Republic of Korea (1973) all passed laws resembling the U.S. Plant Patents Act. Similar to other types of IPRs, industry associations were crucial in ensuring that agricultural inventions were included in the global regulatory framework. In 1911, the Congrès pomologique de France demanded that plant varieties be given extra protection. The issue was again discussed at the International Union of the Horticultural Profession's congresses in Luxemburg (1911), London (1912), and Ghent (1913). According to the International Institute of Agriculture's 1927 Congress, "any grower who engaged in reproduction of those breeds for the purposes of sale to pay a royalty to the producer" needed to be required because the protection of a denomination was insufficient. At its 1931 convention, the International Federation of Breeders of Staple Crops had stated that it hoped new varieties would be granted the same legal status as industrial inventions. At the 1936 International Breeders' Congress in Leeuwarden and the 1937 Conference of the International Organisation of Agricultural Industries, both held in the Netherlands, there were discussions about forming a new organisation to push for the adoption of an international legal framework for the protection of plant varieties. The International Association of Plant Breeders for the Protection of Plant Varieties (ASSINSEL) was established in Amsterdam on November 17, 1938, as a direct consequence of these debates. A three-point resolution was adopted by the inaugural ASSINSEL Congress, which took place in Paris from July 8–9, 1939:

- To reco□nize trademark and appellation filin□s as a form of protection on a global scale (waiting the introduction of a patent);
- To adopt the principle of a license, to be drawn up by ASSINSEL for the purposes of multiplication and sale; and.
- To agree globally that the term "original" refers to seed that is created, offered, or sold by the variety's breeder or by his licensees or successors in title.

These advancements were halted by the Second World War. ASSINSEL passed a resolution at its Semmering Congress in June 1956 calling for an international meeting to establish a global framework for plant variety protection. ASSINSEL had contacted the French government because it had expressed a positive stance. 12 Western European nations received invitations to a diplomatic meeting that will take place in Paris from May 7–11, 1957. The International Convention for the Protection of New Varieties of Plants (UPOV) was enacted in 1961 as a result of these advancements. Section 51(xviii) of the Australian Constitution of 1900 gave Parliament the authority to enact laws pertaining to "trademarks, patents of inventions and designs, and copyrights." Whether agriculture is a suitable subject for intellectual property protection was one of the High Court of Australia's threshold considerations. In The Grain Pool of WA v. The Commonwealth1, the plaintiff contested the constitutionality of the Plant Variety Rights Act 1987 (Cth) and its replacement, the Plant Breeders' Rights Act 1994 (Cth), and the Court helpfully took this question into consideration. Franklin, a protected barley variety, was given to the Grain Pool for the sole purpose of growing trials and malting evaluation. It was claimed that the Grain Pool had produced the barley in Western Australia without authorization, and that commercial negotiations with the agent of the Tasmanian proprietor of the variety had fallen down. In response, the Grain Pool questioned whether the plant variety ri□hts laws qualified as intellectual property under the Australian Constitution. According to the Court, it would be incorrect to consider the legislative granting of monopoly rights in novel plant kinds as falling outside of the "central type" of inventions covered by patents in 1900.

6.1. DEFINITION OF TERMS

The words "local knowledge" and "indigenous knowledge" or "traditional knowledge" were distinguished in an assessment of the role that local knowledge plays in agricultural innovation and food security. It is also helpful to think about how much of this knowledge could be categorised as "intellectual property." These terms became confused during negotiations within the World Intellectual Property Organization's (WIPO) Intergovernmental Committee on Intellectual Property and Genetic Resources, Traditional Knowledge and Folklore (IGC), which is attempting to settle three treaties aimed at recognizing, regulating, and protecting genetic resources, traditional knowledge, and traditional cultural expressions.

6.1.1. Local Knowledge

The information that members of a particular community have acquired and are still acquiring is known as local knowledge. It is a set of facts that pertains to the full range of ideas, opinions, and perceptions that people have about the world. This encompasses how humans measure and observe their environment, solve problems, and verify new knowledge. It encompasses the methods by which information is created, saved, used, and shared. It is:

- Based on experience;
- Often tested over centuries of use;
- Adapted to the local culture and environment;
- Embedded in community practices, institutions, relationships and rituals;
- Held by individuals or communities; and
- Dynamic and changing.

Tribal groups and the original occupants of a region are not the only ones with access to local knowledge. It's not even limited to those living in rural areas. Instead, all communities—rural and urban, settled and nomadic, native and immigrant—have local knowledge. Other concepts that are closely related to, partially overlap with, or even synonymous with local knowledge include indigenous knowledge and traditional knowledge. When it comes to its origin or content, the word "local knowledge" appears to be the least biassed. It incorporates traditional and indigenous knowledge systems because it encompasses a wider range of knowledge systems. Knowledge systems are dynamic, people take in and integrate ideas from a range of sources, and they adjust to changes in their surroundings. Knowledge and access to it, however, are not dispersed equally within or between communities. Individuals may differ in their goals, passions, viewpoints, convictions, and availability of data and resources. Interactions within particular social and agro-ecological environments produce and disseminate knowledge. It has to do with power access and control. Perceptions, access to information, and—most importantly—the value and legitimacy of one's knowledge can all be impacted by differences in social standing. The wisdom that the rural poor, especially women, possess is frequently disregarded. Every culture or community has its own local knowledge, which is held by both young people and seniors. Additionally, distinct types of knowledge are possessed by men and women, farmers and merchants, and educated and illiterate individuals such as:

- The majority of people in a society possess common knowledge, such as the ability to make rice or another basic cuisine;

- Many, but not all, members of the community possess shared knowledge; for example, villagers who keep livestock will have a greater understanding of fundamental animal husbandry than those who do not; and
- Few villagers will become healers, midwives, or blacksmiths, for example, because specialized knowledge is possessed by a select few who may have had specialized training or an apprenticeship.

People's age, gender, occupation, labour division within the family, business, or community, socioeconomic standing, experience, environment, history, etc. all influence the kind of knowled□e they possess. Research and development activities will be si□nificantly impacted by this. The ri□ht people must be found in order to learn what people know. For instance, if boys are in charge of herding, they can be more familiar than their fathers with the locations of the greatest grazing areas. We may only receive a partial picture if we ask the dads to show us good meadows. When the wrong people have been questioned, development professionals may believe that villages don't know much. It is crucial to understand that, like other forms of information, local knowledge is dynamic and ever-evolving as it adjusts to a changing context. It can be challenging to determine whether a technology or practice is indigenous, imported, or a combination of indigenous and introduced elements since local knowledge evolves with time. The latter scenario is more plausible in the majority of circumstances. However, whether a practice is truly local or has already been infused with introduced information is irrelevant for a development effort. Examining what is available within the community is crucial before searching outside of it for technologies and solutions. One can decide what kind of information would be more pertinent to the particular circumstance based on this information. It will probably be a mix of various information kinds and knowled□e sources. Once more, this has si□nificant ramifications for the process of research and development. Documenting the local knowledge that already exists is insufficient. Understandin□ how this knowled□e evolves, adapts, and changes throughout time is equally crucial. It's also important to consider who and how this information is shared both inside and outside the community. Locals have made si□nificant contributions to □lobal knowled□e, such as in the fields of human and veterinary medicine. The environment is □radually changing, and local knowledge is constantly being generated and adjusted. It is deeply entwined with people's cultural beliefs and is transmitted from one generation to the next. The ability of a nation to create and mobilize knowledge capital is just as crucial to sustainable growth in the new global knowledge economy as the availability of financial and physical capital. Local knowled□e is the foundation of any nation's knowledge system. This includes people's abilities, knowledge, and perceptions used to preserve or enhance their standard of living. Many local knowledge systems are in danger of being extinct today. This is due to the fast-paced economic, political, and cultural developments as well as the constantly changing natural surroundings on a worldwide scale. When practices become unsuitable, meet new difficulties, or chan□e too slowly, they cease to exist. However, the introduction of foreign technologies or development philosophies that promise immediate benefits or problem-solving causes many practices to vanish. For individuals who have established and rely on local knowledge for their livelihood, the tragedy of its imminent extinction is particularly poignant. When food is scarce and there are no other options for meeting basic needs, these plants are extremely important for the survival of the impoverished. Furthermore, the loss of experience, technology, artefacts, problem-solving techniques, and abilities may have negative effects on others. People's lives involve local knowledge. In particular, the majority of the impoverished rely on specialized knowledge and abilities that are necessary for survival. Local knowledge is therefore especially pertinent to the following sectors and tactics during the development process:

- Agriculture, including information on planting schedules, intercropping, and crop choices;
- Knowledge of breeding techniques, livestock traits and needs, animal husbandry, ethnic veterinary treatment, and the use of plants to cure common ailments;
- Utilising and managing natural resources, understanding how to control soil fertility, and managing wild species sustainably;
- Medical treatment and understanding of the therapeutic qualities of plants;
- Links between generations are created via community development and shared knowledge; and
- Reduction of poverty and understanding of local resource-based survival techniques.

Traditional methods suggest that technology transfers from supposedly more developed locations are always necessary for development processes. The potential of regional experiences and customs has frequently been overlooked as a result of this practice. At all three stages of the development process, local knowledge is important which include:

- It □oes without sayin□that it is most si□nificant to men and women, young and old, in the local community where those who possess this expertise reside and work;
- In their interactions with local communities, development agents

(CBOs, NGOs, governments, funders, local leaders, and private sector enterprises) must acknowledge, value, and understand local knowledge. Before incorporating it into their methods, they must have a clear understanding of what it is. They must also evaluate it critically in light of their intended goals' utility; and

- Last but not least, local knowledge contributes to global knowledge. It is valuable and relevant in and of itself in this setting. It is possible to adopt and modify local knowledge, transfer it, or conserve it elsewhere.

It is crucial to emphasize, though, that local knowledge is neither unique nor always adequate to address the issues that people face today. Numerous pieces of evidence demonstrate that local actors look for ideas and information wherever they can in order to address their issues and accomplish their objectives. Local knowledge should be viewed as one element of a more intricate innovation system by those engaged in research and development activities with local populations. Therefore, a crucial first step in every research or development effort is a thorough study of the information and knowledge sources that are currently available. By their very nature, these sources can be both official and informal. Community organisations that engage in comparable farming methods, for example, may serve as an unofficial repository of local expertise. Formal knowledge sources would be extension or research centers at the regional or national level. Given their growing significance as information providers, private service providers like neighbourhood seed merchants should be taken into account in this context.

6.1.2. Traditional/Indigenous Knowledge

The information that is represented in the beliefs, abilities, and comprehensions derived from past encounters between people and the physical environments in which they have lived is known as indigenous knowledge systems, or IKSs. Since this knowledge has been handed down through the years, it is firmly embedded in the natural heritage and culture of specific communities and/or cultures. Thus, IKS serves as the foundation for the local knowledge that many indigenous and rural people use on a daily basis and that frequently shapes their choices. The indigenous people who developed these systems have been able to survive thanks to the development of IKS in relation to livelihoods and natural resources. In addition to being represented through storytelling, music, mythology, idiomatic expressions, festivals, cultural norms, beliefs, ceremonies, social regulations, local languages, farming practices, plant species, and even the domestication of animals, IKS is preserved within communities' memories and behaviours. Communities all across the world have used IKS to help them achieve food security. Rural communities' subsistence farmers are continuously battling how climate change is affecting their capacity to provide food security. Many subsistence farmers are depending more and more on different IKS to help them reach their objectives in the fight for food security. These systems differ from local people' cultural knowledge of how to react to early warning indicators of bad weather, the expected effects of wind direction changes, the shape of the crescent moon, seed selection, and animal behaviour. In order to reduce risk and increase crop output, farmers can plan their farming operations based on accurate rainfall likelihood and other meteorological forecasts from IKS. Even while IKS is becoming more and more recognized as a potential means of ensuring food security in rural areas, its application needs to be more methodical if it is to successfully increase agricultural output. Because of this, the usage of IKS is becoming more and more integrated into development agendas that are associated with the work of rural subsistence farmers.

6.1.2.1. Local and Traditional Knowledge Critique

The ideas of "indigenous" and "traditional" knowledge have been linked to a number of issues. The fact that this has come to be linked with the right of peoples to self-determination is a political barrier to their protection. "The protection of cultural and intellectual property is connected fundamentally with the realization of the territorial rights and self-determination of indigenous peoples," according to Erica-Irene Daes's 1993 report, which was written by the chairperson of the Working Group on Indigenous Populations and Special Rapporteur of the UN Sub Commission on Prevention of Discrimination and Protection of Minorities (Daes, 1993). According to Daes, the term "indigenous" accepted the idea of a unique and distinct culture and way of life, founded on enduring customs and knowledge that are essentially tied to a particular region. Le Gall noted that limiting the preservation of traditional knowledge to indigenous works ignores the contributions of majority populations whose members may have entered a nation as slaves or indentured servants. Conflating "local" or "traditional" knowledge with "indigenous knowledge" is difficult because the terms have distinct meanings in developing Asian nations than in settler colonies. As a result, several Asian nations have voiced concerns over the phrase "indigenous people" and thought it was more suitable to use it in relation to the colonial era in Anglo American settler territories.

6.1.3. Legal Transplants

It is noted that in the process of creating legal classifications, communities are expected to meet the expectations of outsiders, particularly lawyers and policymakers, regarding the "authenticity" of their "traditional lifestyles." This is done in an effort to incorporate local, traditional, and indigenous knowledge into international legal instruments. Additionally, it has been argued that calling anything "traditional" when it is really a "admixture of local folk knowledge and extra-local scientific knowledge" is problematic. Some have referred to this as "peasant science." Additionally, it is argued that the term "traditional" does not adequately describe composite agricultural systems (Dove, 2005) that combine market-oriented cash crops with subsistence-oriented swiddens, and the term "indigenous knowledge" does not adequately convey the historical movement and exchange of plant genetic resources among various regions of the world that underpin local farmer knowledge and practices (Dove, 2000). As early as its third session, the IGC saw a tendency "to leave specific determinations of the boundaries of protectable subject matter up to domestic authorities, and for terminology at the international level to be used more to express a common policy direction." Therefore, it noted that a broad definition of the scope of intellectual property (IP) protection, particularly on a global scale, can be separated from the more specific standards that are created and implemented on a case-by-case basis at the national or regional level. Drahos refers to this as the "tragedy of commodification." A more basic issue with attempts to incorporate traditional, indigenous, and local knowledge into the intellectual property paradigm is that it reflects the western idea of separating the object of protection from the environment in which it evolved. In addition to being seen as a colonialist form of legal transplantation from industrialised to developing nations, the WIPO IGC's attempts to fit these categories of knowledge within treaties that allow for their exploitation are also thought to incite intra-community conflicts over benefit sharing (Forsyth, 2015). Bragdon points out that genetic resources in general, rather than the creative process where all germplasm, whether traditional or modern, "is treated as a potential input for direct use or further improvement," are the main topic of discussion in the IGC about agricultural improvement by farmers. The customary law of traditional societies is consistently eliminated as being unsuitable or ambiguous as a result of the legal transplantations required to include traditional knowledge into the process of creating international treaties.

6.1.4. Plant Variety Rights

Plant breeders' rights (PBR), also called plant variety rights (PVR), are rights that are granted to the breeder of a new plant variety in some places. These rights give the breeder exclusive control over the new variety's harvested material (cut flowers, fruit, and foliage) and propagating material (seed, cuttings, divisions, and tissue culture) for a number of years. With these rights, the breeder has the option of licensing the variety to third parties or becoming the variety's sole marketer. A variety needs to be novel, unique, consistent, and steady in order to be eligible for these exclusive rights. A range is:

- **New:** It has not been commercialized for more than one year in the country of protection;
- **Distinct:** It differs from all other known varieties by one or more important botanical characteristics, such as height, maturity, color, etc.;
- **Uniform:** The plant characteristics are consistent from plant to plant within the variety;
- **Stable:** The plant characteristics are genetically fixed and therefore remain the same from generation to generation, or after a cycle of reproduction in the case of hybrid varieties.

The International Convention for the Protection of Plant Varieties (UPOV) serves as the model for specialized (sui generis) laws that protect plant varieties in the majority of nations. A number of restrictions were included in the 1961 UPOV Convention, including the need that nations choose whether to use patents or sui generis legislation to protect UPOV. The USA, which already had a patent system in place to deal with plants, was essentially left out of this restriction. In 1978, a clause allowing farmers to harvest protected seed for further planting or trade with other farms was added to the Convention. The farmers' privilege was limited to harvesting seed in reasonable quantities for re-use on a farmer's own property in a final revision in 1991 that made the UPOV convention appealing to the USA and allowed for double protection of plant varieties through patenting and sui generis protection. Additionally, varieties that were "essentially derived" from protected kinds were protected under the 1991 version of UPOV. A "breeder" or individuals claiming via a breeder—defined in Article 1(iv) of the UPOV Convention as the person who bred, discovered, or produced a variety—are granted protection under UPOV. Generally speaking, "breeding" refers to both the discovery of a plant and its application in selective propagation to accomplish a goal. In general, the following actions or licenses are granted to the plant breeder in connection with propagating the variety's material under plant variety rights legislation:

- Produce or reproduce the material;
- Condition the material for the purpose of propagation;
- Offer the material for sale;
- Sell the material;
- Import the material; and
- Export the material;

According to laws based on the UPOV Convention, plant variety rights last for 20 years for any other variety and 25 years for trees and vines. Following a registration procedure, plant variety protection is established. If a plant variety has a breeder, is unique, consistent, and stable, and hasn't been exploited or has only recently been, it is deemed registrable. If a plant variety can be easily identified from any other variation whose presence is well known, it is said to be distinct. It is uniform if its pertinent qualities during propagation are uniform, subject to the variance that may be anticipated from the specific features of its propagation. If a plant variety's pertinent traits hold true after multiple propagations, it is said to be stable. If a plant variety or its propagating material is not sold to another individual by the breeder or with their consent, it is considered not to have been exploited. The grantee of plant variety rights is generally required by legislation based on the UPOV Convention to take all reasonable measures to guarantee reasonable public access to the plant variety. If propagating material of acceptable quality is made available to the general public at affordable costs or as gifts in enough numbers to satisfy demand, this criteria is deemed to be met. On fair terms and conditions, a suitable individual may be granted a license to sell or create propagating material of that type of plant. In general, a plant variety that is not directly used as a consumer product is exempt from the requirement to get a compulsory license. Under Article 27.3(b) of the TRIPS Agreement, signatories are required to safeguard plant types. Although nations can choose to protect plant varieties through patents, sui generis protection, or a combination of the two, most have preferred to comply by adopting the 1991 Act of the UPOV Convention since the TRIPS Agreement began in 1995. As of May 2, 2019, there are 75 signatories to the UPOV Convention, 41 of whom joined after January 1, 1995. Only a small number of nations have embraced alternatives to UPOV, despite a plethora of comments and suggestions for the adoption of alternative sui generis models. This is especially true for India, Malaysia, and Thailand, which have incorporated benefit-sharin□clauses inspired by the CBD into PVP. One of the reasons why nations have tended to embrace UPOV 1991 instead of creating a sui generis alternative is because the USA and EU have included an obligation to join UPOV 1991 in the IPR chapters of the free trade agreements they have signed with their bilateral partners since the 1990s.

6.1.5. Plants as Intellectual Property

When evaluating the general need for the protection of developed plant varieties as a type of intellectual property, scientists and lawyers have held differing views. At the moment, concepts that can be proven to be original and unexplored at the time of their legal claim as intellectual property are protected by intellectual property rights. However, throughout the history of intellectual property law, both domestically and internationally, this meaning of uniqueness has remained ambiguous. The legal conditions for the first plant patent are different from the anticipated future modifications to the legal protection of intellectual property connected to plants. These legislation' proponents acknowledge the general necessity for funding research and development. For instance, agricultural research and development has been identified as a particularly challenging undertaking, given the pressing concerns over the ability to feed a growing world population in a sustainable manner. Conversely, other people think that a more varied strategy is needed than simply enforcing rules pertaining to intellectual property rights on novel plant kinds. According to this counterargument, the nature of intellectual property and its preservation are influenced by many social, cultural, and economic considerations. One particular issue raised by this argument is how seeds are obtained in various domestic and foreign locations. Supporters of this viewpoint contend that in order for intellectual property rights rules to effectively safeguard the rights of plant breeders, they must take into account the fact that this process is very ephemeral and can change significantly over time. In the US, new laws have been put into place as a result of this dispute over control over seeds. One national initiative, the Open-Source Seed Initiative (OSSI), was launched in the United States and is the first of its type to base its approach to plant breeders' rights on the processes used by freely sourced software. As worries about the usage of open source technology within a legal framework have grown, there has been more discussion about this strategy. According to some, OSSI has severely restricted plant breeders' access to new plant types' intellectual property rights. As a result, there have been assertions that research and development funding in this field will likewise decrease.

6.1.6. Seed Sovereignty

The right of farmers to save, breed, and trade seeds, as well as to have access to a variety of open-source seeds that are viable and not patented, genetically modified, owned, or controlled by new seed goliaths, is known as seed sovereignty. It generally stems from the idea that in order to improve agricultural biodiversity, resilience, and food security, communities ought to be in charge of their own seed stock. The significance of seed sovereignty lies in:

- It is the cornerstone of preserving both food security and biodiversity;
- The cost of buying seeds can be decreased by producing high-quality open-source seeds cheaply and simply;
- Permits farmers to grow their own seeds (seed self-reliance) and freely trade and sell seeds with one another;

- Allows farmers to choose seeds that are appropriate for their climate. One can select seeds of plants that are grown in their garden with these particular characteristics if they desire large fruits that are not affected by pests; and
- It makes it possible to save priceless native or traditional vegetable seed variations for upcoming generations.

Given the si□nificance of seed savin□ for seed soverei□nty, this concept is strongly tied to concerns about intellectual property rights, especially those pertaining to the patenting of plant genetics. Advocates contend that the process of preserving ancient plant kinds ought to be legally protected for both farmers and individuals. In addition, proponents of seed sovereignty contend that environmental and food security reasons justify the protection of seed preservin□ Given the cultural si□nificance of particular seeds and plant kinds, particularly among indigenous groups, some activists contend that seed sovereignty is crucial. Because it emphasises boosting food security for all communities, seed sovereignty has close linkages to the food justice and food sovereignty movements.

6.2. PATENTING AGRICULTURE

PBRs, a comparatively feeble type of legal protection that predominated in the majority of developed countries, were the sole intellectual property protection available for crop plants until around 1980. When a seed owner received a PBR certificate from the nation's department or ministry of agriculture, rivals were prohibited from selling seeds or breeding material from the owner's particular seed variety. In their own breeding operations, competitors were permitted to exploit the protected types as sources of subsequent seed variety, nevertheless, thanks to the PBR. Regular patents, which are far more limited than PBRs, were later allowed in the US for living things like seeds and plants, as well as for genes and a number of other biological plant components and diagnostic items. The burgeoning biotechnology sector has been facilitated by these expanded intellectual property rights. For a set number of years, a patent is a statutory license that a government grants to an inventor and those people who derive their rights from the inventor to prevent others from producing, using, or selling a patented product or from employing a patented method or process. The individual who is eligible to file for a patent and who satisfies the necessary registration conditions is granted patent rights by statute as a matter of right. Typically, the protection that comes with registering a patent has a time restriction. For instance, the period of protection is 20 years under s.25 of the UK Patents Act 1977. The patented invention is considered to be in the public domain at the conclusion of the protection period, meaning that anybody can use it. Generally speaking, an invention is an idea that makes it possible to solve a particular technological challenge. The inventor or his successor in title is often the one applying for an invention's protection. Most legal systems require an innovation to be the following in order to be protected by a patent:

- New;
- Involve an inventive step;
- Be industrially applicable; and
- Not be a category of excluded invention.

Legislation and case decisions may declare something that would have been seen as a discovery to be an invention since intellectual property rules are derived from legislation that is interpreted by judges. In its Biotechnology Directive, for instance, the European Parliament stipulates that biological material that is separated from its natural habitat or created through a technical procedure is considered an invention, regardless of whether it was naturally occurring. In Diamond v. Chakrabarty, the US Supreme Court adopted a similar stance, holding that a microbe that was genetically modified to break down crude oil was an innovation. The engineering of desired genetic features from beneficial indigenous species has been made possible by the current biotechnology revolution. Many desired features may now be quickly introduced into plants thanks to genetic engineering. These include agronomic traits like nitrogen fixation and utilization, restricted branching, environmental stress tolerance, male and/or seed sterility for hybrid systems, post-harvest traits like delayed ripening of fruits prone to spoiling, and pest control traits like insect, virus, and nematode resistance as well as herbicide tolerance. Output traits include plant colour and vitamin enrichment. The creation of several enabling and transformative technologies has made it feasible to produce transgenic plants. Due to the positive rulings of US and European courts, these methods, together with the genetic location of advantageous plant features, are now protected by patents. This has an impact on agricultural innovation in that unapproved individuals are held accountable for patent violation when they grow, distribute, and market crops that contain patented genes. The widest range of agricultural patents is allowed in the United States. For entire plant lines, including particular herbicide-resistant rice lines, it has routinely granted patents. In the majority of countries, where PBR protection is the only option for plant lines, such variations are most likely not patentable. The Supreme Court is now examining Pioneer v. J.E.M. Agric. Supply, an appellate case that established the patentability of plant types in the United States early in 2000. These patents usually cover the plant's offspring as well as its seeds. The protected seed's usage in U.S. research for developing-world purposes will be limited because the claims are obviously intended to prevent other breeders from using it as breeding material. On groups of plant kinds, there are also very wide U.S. and European patents. For example, the U.S. Agracetus patents aim to cover all transgenic soybeans and cotton. If these patents are legitimate, Monsanto, the company that purchased Agracetus, may be able to control all transgenic versions of these crops. However, patents pertaining to certain technical processes utilised in a□ricultural □enetic en□ineerin□are of the utmost si□nificance to plant breeders. For instance, the majority of the technology used to produce hybrid rice was created in China, where a si□nificant amount of the nation's rice is produced using hybrid seed. The China National Seed Corporation patented several parts of the technique in the US, but their early patents are no longer valid. Breeders are prevented from using research instruments that could be helpful in creating new crop types by these patents. Other methods of producing hybrid seed have also been the subject of patents. A U.S. patent for the gene gun, one of the most popular methods for introducing genes into plants, may further restrict research. Cornell University received it and granted DuPont a license. The 35S promoter, a section of DNA that is frequently inserted with a plant gene to increase its expression, is likewise patented by Monsanto. It will be far more difficult and costly for breeders to create superior seeds if they are unable to use such technologies or require licenses to do so.

Nowadays, it is common practice to patent genes themselves, usually with claims that cover the isolated gene, different gene-containing structures, plants that have been altered by those constructs, and the seed and offspring of such plants. Since naturally occurring plants that carry a certain gene are not innovative, neither breeding nor the patent apply to them. However, it might be forbidden to use the gene, its constructions, seeds, or offspring for any other purpose. For instance, the Xa21 Kinase □ene, which confers disease resistance on □rains, is patented by the University of California. The □ene was identified thanks in large part to work done at IRRI, and the university made arrangements to safeguard IRRI's usage of the gene. Some other genes, on the other hand, are safely in private hands with no promise to make them publicly available. This is true for certain patents that involve introducing genes encoding viral coat proteins—which provide resistance to plant viruses—into plants. Many patents for Bacillus thuringiensis (Bt) technology, in which bacterial genes are put into plants to produce toxic proteins that kill insects, also fall under this category. Hundreds of frequently overlapping patent rights have resulted from the loose issuance of Bt claims, and they have been the focus of intense litigation. For instance, Bt-transformed maize has been claimed by at least four distinct corporations. It is nearly hard for a researcher to navigate this maze of patents. Usually, trade secrecy procedures are used to secure genomic data. In this method, access is granted only under predetermined conditions, which may include a compensation mechanism, by a corporation that develops a sizable database or genome map. Additionally, this paradigm serves as the foundation for significant global nonprofit collaboration. For instance, Japan, Korea, China, the United States, the European Union, and the Rockefeller Foundation are all participating in a global genome sequencing effort through the International Rice Genome Sequence Working Group because rice is crucial to the world's poor and its genome is smaller than that of some other cereals. The participants have consented not to submit patent applications for the sequences, and the information will be added to public databases. Monsanto has created its own sequencing and consented to release its genomic rice data for public breeding in underdeveloped countries. Although Syngenta and Myriad Genetics finished a rice sequence in January and have committed to sharing knowledge and technology to advance subsistence farming globally, they are not making their sequences publicly available. Furthermore, it's unclear if other genomes or the genomes of significant pathogens will be as easily accessible, and many of the crucial rice genes might be patented.

There is a huge concentration of agricultural biotechnology that coincides with this patenting trend. A significant portion of the agricultural patent portfolio is currently held by five major corporations: Aventis, Dow Chemical, DuPont, Monsanto, and Syngenta. These companies have merged with chemical and pharmaceutical companies to gain access to production capacity and chemical markets, acquired seed companies worldwide to enhance their ability to market new products, and acquired smaller biotechnology companies to acquire the technologies developed by those companies. They have accumulated extensive portfolios of intellectual property in the process. The amount of agricultural research is decreasing as the industry becomes more concentrated. Recent consumer and environmental criticism of bioengineered foods may have contributed to the decrease, although industrial consolidation may also have reduced incentives. Some of these big companies have actually started to show interest in emerging global markets in recent years. The interest is greatest in soybeans and the major grains (rice, wheat, and maize), where there are sizable developing-world markets and significant export potential. However, it also includes rice, whose seed was previously thought to be an essentially noncommercial product that was provided by public institutions for free or at a reduced cost. Better cultivars like IR-16 and IR-64 were created at IRRI with donor funding during the Green Revolution. The institute freely distributed novel breeding materials and new kinds to national research facilities in the main East Asian countries. They then released the produced cultivars to national systems for production and farmer distribution after further breeding them to suit local growing circumstances. In Asia, these public types are the most common, but businesses are starting to enter the market. Research initiatives have been created in India by Pioneer, which is currently owned by DuPont. There, private hybrid rice breeders like Mahyco have also surfaced. Monsanto and the Indian Institute of Science have collaborated on scientific projects. In Japan, tobacco started to show interest in rice seeds. Additionally, before to being purchased by Monsanto, Cargill's developing-world divisions had already started a hybrid rice breeding program. These and other agricultural companies are attempting to defend their intellectual property positions in important developing countries, such as China and Brazil, according to global patent searches. Important research methods, instruments, and gene constructions are probably going to be patented in at least some of these countries, even if they do not offer the same range of legal protections that are available in the US.

The expansion of si□nificant commercial markets in emer□in□ countries is reflected in the private sector's interest in supplyin□ rice seed to these countries. The value of the rice produced in the two major Asian markets is si□nificantly □reater than that of the corn crop □rown in the United States, which has sparked a great deal of private research. Since harvested rice may typically be utilised as seeds, this does not instantly turn into a seed market. Investment from the private sector will rely on a proprietary position of some kind, such as successful hybrids or plants that are shielded by intellectual property rights or by a "terminator" technology that renders rice plants sterile. Although reaching this position could be challenging, businesses are willing to try because of the promise of Asian rice. Additionally, the companies have a financial stake in chemical marketing. A company can expand the herbicide's market by introducing the genes required for herbicide resistance into national crop lines. Herbicide intellectual property rights are already present in China. India's rules mandate that full patents be awarded by 2005, and it has granted exclusive marketing rights. Multinational corporations will most likely introduce seeds that are superior to the current crop when they join areas like the Asian rice seed market. This is good. Furthermore, a number of scientists contend that using plants resistant to herbicides is a more environmentally friendly method of controlling weeds than using other methods. However, it is likely that private sector seeds will only be created for the larger commercial markets; it will be a long time before the private sector helps poor farmers or enhances small crops. More importantly, there is a very real chance that the multinational industry will have an oligopoly or monopoly on transgenic seeds due to patent rights and the small number of large companies, excluding rivals and even the government. As a result, prices will likewise be higher than they are now. Lastly, the public sector may not be able to acquire patented technologies or employ protected varieties for research in order to find new applications for the smaller crop or subsistence farmer, or it may be extremely costly or difficult to do so.

6.2.1. Infringement of Agricultural Patents

Unauthorized use of patented content could make the user liable for the patent. Potential defendants in the agriculture industry include farmers who cultivate seed containing patented DNA, traders in proprietary seed, and research institutions that use or distribute seed containing patented DNA. Some of the more notable infringement instances are listed below.

6.2.1.1. 1Monsanto Co. v. Scruggs

The US patent of a gene that gave glyphosate resistance to crops like canola, soybeans, and cotton that had the patented gene introduced and made them resistant to Monsanto's glyphosate herbicide "Round Up" was at issue in the case of Monsanto Co. v. Scruggs7. In order for seed companies to sell seed that contained Monsanto's patented genes to growers, they had to first sign a technology license agreement that restricted buyers to growing a single commercial crop. This meant that growers could not save seed from a harvested crop for replanting the following growing season. Monsanto had licensed this Round Up Ready ("RuR") technology to seed companies. Mitchell Scruggs bought a limited amount of RuR soybeans from a Memphis seed company without a technology licensing agreement. About 10 acres of soybeans might be planted with the seed. Mr Scruggs kept the soybean seed from those 10 acres after the autumn harvest; he cleaned it and stored it for use in the 1997 growing season. Scruggs was able to save enough RuR soybean seed to plant over 8000 acres by the year 2000 by preserving seed from every crop season that followed. Monsanto asserted that Scruggs had violated its patent. Scruggs argued that the patent did not cover Monsanto's biotechnology or the plants grown in their fields, but the Court ruled in favor of Monsanto. The fact that Scruggs had personally acquired Monsanto's patented technology clearly put him at risk.

6.2.1.2. Monsanto Co. v. Schmeiser

A Canadian lawsuit involving Monsanto Canada, Inc. and a farmer named Percy Schmeiser, who had never bought the patented technology, is more important when evaluating the effects of patenting on agriculture. Monsanto commercialized "Roundup Ready Canola" in Canada after obtaining a patent there in 1993 for glyphosate-resistant plants. In Saskatchewan, Schmeiser farmed canola for a living. He didn't have a license to plant RuR Canola, and he had never bought any. However, experiments conducted in 1998 showed that RuR plants accounted for 95%–98% of his 1000 acres of canola production. It's unclear where the plants came from. They might have originated from RuR seed that drifted into or close to Schmeiser's property. Then, when Schmeiser sprayed Roundup herbicide in the ditches along the road that borders four of his fields and around the power poles, the plants that survived were collected. By sowing glyphosate-resistant seeds, Schmeiser was allegedly using, reproducing, and creating genes, cells, plants, and seeds that included the genes and cells claimed in the plaintiffs' patent, according to Monsanto's lawsuit alleging patent infringement. Schmeiser claimed during the case trial that Monsanto had lost or relinquished their right to assert an exclusive patent on the gene since they had not controlled its spread and had no intention of doing so by releasing it into the environment without any restrictions.8. Schmeiser went on to claim that the patent was null and void due to:

- The claimed invention is self-propagating and may spread without human assistance; it is a life form meant for human consumption and is not the appropriate subject matter for a patent;
- The patent was obtained illegally in order to create a noxious plant that would naturally spread to innocent persons' lands and trap them with allegations of nuisance patent infringement; and
- The plaintiffs would effectively obtain a patent for a plant if infringement is found, which is argued to be impossible in Canada due to the Plant Breeders' Rights Act (PBRA), which protects new plant varieties and upholds a farmer's right to save and reuse seed.

Each of these arguments was dismissed by the trial judge, who concluded that Schmeiser had violated Monsanto's patent. He maintained that the creation of the gene and the procedure for inserting it could still be registered as an invention under the Canadian Patent Act even though replication of the gene might happen naturally, without human intervention, after it was inserted in the original plant cells and plants that were grown for seed. The Trial Judge noted that Schmeiser had cultivated RuR-tolerant canola from seed. According to his ruling, the process of growing the seed, replicating the copyrighted gene and cell, and selling the harvested crop amounted to stealing the core of Monsanto's invention, exploiting it without authorization, and violating the patent. After a failed appeal, the Federal Court of Appeal determined that Schmeiser had engaged in an unauthorized use of the proprietary material when he saved and planted seed, then harvested and sold plants that included the patented cells and genes.

6.2.1.3. Monsanto Co. V. McFarlin

In Monsanto Co. v. McFarling, the US Federal Circuit Court examined the connection between patents and plant variety rights. In order to sell its patented seeds, Monsanto had mandated that sellers obtain a "Technology Agreement" from buyers. This agreement stated that the buyer would not "save any crop produced from this seed for replanting, or supply saved seeds to anyone for replanting," and that the seeds would only be used "for planting a commercial crop only in a single season." In 1997 and 1998, Mississippi farmer Mr. McFarling signed the Technology Agreement and bought Roundup Ready soybean seed. One season, he preserved 1500 bushels of the patented soybeans from his harvest; the following season, he planted them as seed rather than selling them as a crop. In the subsequent growing season, he carried out this task once more. The Roundup Ready seed's genetic alterations were preserved in this conserved seed. Although Mr. McFarling did not contest that he had broken the terms of the Technology Agreement, he argued that the clause in the contract that forbade him from using the patented seed to create new seed for planting—when he only produced enough new seed for his own use the next season—violated the US Plant Variety Protection Act's (PVPA) seed saving clause, which allowed farmers to preserve seeds of plants that were registered under the Act. The Court decided against Mr. McFarling in Pioneer Hi-Bred International Inc. v. J.E.M. Ag Supply Inc., refusing to use the PVPA to restrict the patent law.

6.2.1.4. Monsanto Technology LLC v Cefetra BV

Even if there isn't a patent in the exporter's nation, importing genetic material that has been patented can nevertheless result in patent infringement. Farmers and agricultural product exporters may find this to be a cause for concern. The patented gene must, however, be able to carry out its patented function in the exported material, according to a collection of relevant European instances. In 2008, Monsanto filed infringement lawsuits against Argentine importers of soy meal made from its proprietary glyphosate-resistant soy. In the case of Monsanto Technology LLC v. Cefetra BV and the State of Argentina in the Netherlands, Monsanto requested an injunction that would forbid patent infringement in all European nations. The importer denied infringement, claiming that the patented DNA was dead material that was unable to express the required enzyme since soy beans were processed to make the meal. The European Court of Justice ruled in favor of this interpretation. The UK High Court had adopted a similar stance.

6.2.1.5. Organic Seed Growers & Trade Ass'n v. Monsanto Co.

According to Monsanto's website, "we have only filed suit against farmers 147 times in the United States since 1997." When you take into account that we sell seed to over 350,000 American farmers annually, it implies that this is "really a small number." The Organic Seed Growers and Trade Association led a coalition of 38 farmers, seed sellers, and agricultural organisations that sought declaratory judgements of non-infringement and invalidity in the District Court for the Southern District of New York15 regarding 23 patents owned by Monsanto Co. and Monsanto Technology, LLC despite the slight risk of litigation. As producers, seed companies, and agricultural associations that cultivate, use, or market conventional seeds—many of which are certified organic—the plaintiffs stated that they had no intention of using or marketing transgenic seed that integrated Monsanto's technology. Their main worry was that if their crops were tainted by transgenic seed, the company that produced the tainted seed may falsely accuse them of patent infringement due to Monsanto's patent enforcement policy. On the grounds that there was no justiciable dispute between the parties, the District Court denied the plaintiffs' request for a declaration. It stated that there was no proof that Monsanto had filed a lawsuit against unintentional consumers of copyrighted seed, nor was there any proof that any of the plaintiffs had ever been threatened with patent infringement by Monsanto or had been contaminated by the company's seed. The plaintiffs requested an express commitment from Monsanto not to sue shortly after they started the case. "It has never been, nor will it be Monsanto policy to exercise its patent rights where trace amounts of our patented seeds or traits are present in farmer's fields as a result of inadvertent means," the company said on its website, which the plaintiffs accessed while refusing to engage in such an undertaking. When the District Court decided that there was no impending disagreement between the parties, it also took these representations into consideration. The Court of Appeals' conclusion that there was no justiciable dispute between the parties was influenced equally by these factors. Certiorari to allow an appeal was denied by the Supreme Court.

6.2.1.6. Marsh v Baxter

Note that the different decisions in Organic Seed Growers & Trade Ass'n v. Monsanto Co. only applied to a dispute in the United States. In that case, the District Court had observed that it was inevitable that traces of wind-blown pollen or seeds from genetically modified crops or other sources would contaminate conventional crops. Monsanto's admission that conventional crops "may inadvertently contain traces of Monsanto biotech genes (because, for example, some transgenic seed or pollen blew onto the grower's land)" and that they may be subject to "cross-pollination from nearby fields where biotech crops are grown" was also mentioned by the Court of Appeals. The Court of Appeals cited a study that found that Monsanto's Roundup resistant feature had infiltrated the vast majority of traditional seed samples, even though strict precautions were taken to avoid any mixing of modified and normal seed crops. According to the Court of Appeals, a grower who harvests, uses, or sells infected products runs the danger of being held liable for infringement as it is difficult to distinguish genetically modified seeds from traditional seeds. At the very least, "both parties seem to concede that using or selling patented seeds without a license is potentially infringing activity," the Court of Appeals noted. Therefore, regardless of whether the alleged infringer intended to profit from the patented technology, the court assumed for the purposes of the appeal before it that "using or selling windblown seeds would infringe any patents covering those seeds" (without making a decision). In Western Australia, a disagreement between nearby farmers brought up the larger topic of unintentional contamination. Steve Marsh and his spouse had signed an agreement with the National Association of Sustainable Agriculture (Australia) Ltd. (NASAA) to have their farm, Eagle Rest, certified organic. On the other side of a 20-meter road, another property run by Michael Baxter shared a boundary with theirs. He informed Baxter of his intentions to plant Monsanto's Roundup Ready (RuR) genetically modified canola in the paddocks of his farm that were next to Eagle Rest during the beginning of 2010. Marsh personally presented a "Notice of Intention to Take Legal Action" to Baxter on September 29, 2010, which included the following information:

- Non-GM farmers, especially those who have been certified as or□anic (or sustainable) farms, could suffer catastrophic financial losses as a result of the usa□e of □enetically modified or□anisms in farming, such as GM canola seed (GMOs); and
- When GMOs contaminate a non-GM farm or non GM farm production cycle, the main reason why Organic Farmers suffer commercial losses is because they lose the price premiums associated with selling their produce and/or have their accreditation as Organic Farmers revoked.

Notices declaring Eagle Rest was a "GMO Free Area" were issued by Marsh in the local newspapers on October 25, 2010. Baxter harvested the genetically modified canola by swathin□it in November 2010. Cuttin□the immature canola plant near its root was necessary to achieve this. After a few weeks of standing in the paddock to mature, the swaths were processed by a header to extract the ripened seeds from each swathe. Baxter had never swathed his canola rather than harvested it directly before. Around 245 GM canola swathes were delivered to the Marshes' farm in late November or early December 2010 after being carried by the wind. In accordance with NASAA Standard 3.2.9, which states that "Or□anic certification shall be withdrawn where NASAA considers there is an unacceptable risk of contamination from [□enetically modified organisms] or their derivatives," the agency decertified about 70% of Eagle Rest after inspecting the GM canola swathes. This was due to an assessment that the RR canola swathes and seed pods found on Eagle Rest posed a "unacceptable risk" of contamination. The Marshes filed a $AU85,000 lawsuit against Baxter, claiming negligence and private nuisance as the two grounds of action, for the damage resulting from 70% of Eagle Rest losing its NASAA/NCO accreditation. The Court of Appeal of Western Australia and the Supreme Court of Western Australia25 rejected these actions. The negligence suit "traverses into legally unchartered territory," the trial judge noted. He claimed that because the law was reluctant to broaden the scope of situations in which economic damage may be recovered, the claimed duty was new and presented a conceptual challenge. The judge remarked that the Marshes' claimed obligation to "make sure the Marshes did not suffer loss" was unequivocal and excessively high given the circumstances of broad-acre farming that was subject to unpredictable seasonal weather. They could have had a better chance of success if a duty of care had been more precisely defined. From a causal standpoint, for instance, the Marshes' actual complaint was about Baxter's decision to harvest via swathing rather than his decision to cultivate genetically modified canola; therefore, the purported duty of care may have been connected to Baxter's harvesting method choice. On the other hand, it had not been demonstrated that Baxter had grown or swabbed RR canola carelessly. By a vote of 2:1, the Western Australian Court of case denied the case and maintained the lower court's ruling.31 "That in the particular circumstances of this case, the GM plant material that landed on the appellants' farm posed no risk of any genetic trait transfer to any species of crop or produce on the appellants' land" was the unanimous ruling of the majority judges.32 They observed that only eight volunteer GM canola plants were ever found in the next planting season, despite the fact that about 245 swaths had invaded the appellants' land. Mr. Marsh had identified and removed these plants, "presumably before they had set seed."

6.2.2. Patenting of Plant Breeding Methods

It would have been assumed that plant breeding techniques would not be granted patent protection due to the European Patent Law's exclusion of "essentially biological processes for the production of plants or animals," which are defined in Article 2.2 of the Biotechnology Directive as consisting "entirely of natural phenomena such as crossing or selection." However, the EPO Enlarged Board of Appeal recently tested this in two rulings. One was on the potential patentability of a method that involves broccoli crossover and selection. A related invention involving tomato crossover and selection was the subject of another referral. Plant Bioscience Ltd. (Norwich, UK) submitted the broccoli patent application for a "method for selective increase of the anticarcinogenic glucosinolates in brassica species" (see Note 11). The Israeli Ministry of Agriculture submitted a tomato patent application for a "method for breeding tomatoes having reduced water content and product of the method." Interested parties contested both of the patent applications. The Technical Board of Appeal of the EPO heard these objections and assigned several questions to the EBA for resolution.

The enquiries concerning the broccoli patent were:

- Is a non-microbiological method of producing plants that involves crossing and selecting plants exempt from Article 53(b) EPC just because it incorporates an additional technical feature as a step or as a component of any of the crossing and selection steps?
- What are the pertinent standards for differentiating between non-microbiological plant production processes that are not protected by patent under Article 53(b) EPC and those that are, if Question 1 is answered ne□atively? Specifically, does it matter where the claimed invention's core is located and/or if the technical feature adds anything to the claimed invention that goes beyond the obvious?

The questions raised in respect of the tomato referral were:

- Does the exclusion of Article 53(b) EPC apply to a non-microbiological plant production process that involves crossing and selecting plants only if the procedures are representative of and consistent with natural occurrences that may happen naturally without human interference?
- If the answer to Question 1 is negative, does a non-microbiological method of producing plants that involves crossing and selecting plants get around the exclusion of Article 53(b) EPC just because it includes a technical feature in any of the crossing and selection steps?
- What are the pertinent standards for differentiating between non-microbiological plant production techniques that are not protected by patent under Article 53(b) EPC and those that are, if Question 2 is answered ne□atively? Specifically, does it matter where the claimed invention's core is located and/or if the technical feature adds anything to the claimed invention that goes beyond the obvious?

The EBA answered the questions as follows:

- Since it is "essentially biological" in the sense of Article 53(b) EPC, a non-microbiological method of producing plants that includes or consists of the steps of sexually crossing the entire genome of plants and then selecting plants is, in theory, not patentable.

- This process is not exempt from Article 53(b) EPC just because it includes a technical step as an additional step or as a component of any crossing and selection step that facilitates or aids in the execution of the steps for sexually crossing the entire plant genome or for subsequent plant selection.
- However, under Article 53(b) EPC, a process is not excluded from patentability if it includes, within the steps of sexual crossing and selectin□, a technical step that, by itself, introduces or modifies a trait into the genome of the plant produced, provided that the introduction or modification of that trait is not the result of the mixin□ of the genes of the plants chosen for sexual crossing.
- It is irrelevant whether a technical step is a new or known measure, whether it is trivial or a fundamental alteration of a known process, whether it does or could occur in nature, or whether the essence of the invention lies in it when determining whether such a process is excluded from patentability as being "essentially biological" under Article 53(b) EPC.

The following factors were listed as pertinent to assessing whether a process is not fundamentally biological, and the EBA recognized them from the jurisprudence:

- The totality of human intervention and its impact on the result achieved is to be determined.
- This has to be judged on the basis of the essence of the invention.
- The impact must be decisive.
- The contribution must go beyond a trivial level.
- The totality and the sequence of the specified operations must neither occur in nature nor correspond to the classical breeders' processes.
- The required fundamental alteration of the character of a known process for the production of plants may lie either in the features of the process, i.e., in its constituent parts, or in the special sequence of the process steps, if a multistep process is claimed.

During the hearings, there was a contention that the terms "crossing" and "selection" should only refer to natural processes. Specifically, the term "selection" exclusively referred to natural selection—which is uncontrollable by humans—that determines which plants survive in the wild, rather than selection conducted by humans during a breeding process. The EBA decided that, in order to apply the principles of treaty interpretation, a term's meaning could not be determined solely by semantics; rather, it had to be interpreted in good faith and in line with the terms' typical meaning in their context. Therefore, it was noted that a definition that totally i□nored the fact that the phrases "crossing" and "selection" in the EPC's provisions refer to actions taken by the breeder, and that the context of these terms is provided by the techniques used to produce plants. These are distinguished by the breeder's intervention in the processes to produce the intended outcome. Therefore, crossover and selection are method steps that typically require human intervention rather than being natural phenomena in that environment.

6.2.3. Patentable Inventions Technical Requirement

The fundamental question of what botanical discoveries qualify as patentable inventions for the purposes of patent law is brought up by the EBA's Broccoli and Tomato rulings. Depending on the current country patent law, the response to this issue will vary. The most lenient patent laws are thought to be found in the USA. In Diamond v. Chakrabarty, decided in 1980, the Supreme Court had to decide whether a genetically modified microbe that could degrade several components of crude oil may be patented. In that instance, the application was denied by the patent examiner for two reasons:

- That micro-organisms are 'products of nature' and
- That as living things they were not patentable subject matter under the US patent law.

These worries were dismissed by the Supreme Court, which famously cited Congress's goal for the patent system to "include anything under the sun that is made by man" as patentable. The Court decided that the microorganism was eligible as patentable subject matter as a result. The patent claim in question, the Court added, "was not to a hitherto unknown natural phenomenon, but to a non-naturally occurring manufacture or composition of matter—a product of human ingenuity." To put it another way, a biological innovation could not be patented without some human intervention. The requirement that an invention be "technical" in nature is the main focus of the European Patent Office. Patentable biotechnological inventions are those that address the following, according to Rule 27 of the Convention on the Grant of European Patents' Implementing Regulations:

- Biological substance that has been separated from its native habitat or created by a technical procedure, even if it was naturally occurring before;
- Plants or animals if the invention's technological viability is not limited to a certain type of plant or animal; and
- Anything other than a plant or animal variety that is produced throu□h a microbiolo□ical or other scientific technique.

In determining whether plant breeding techniques were patentable, the EBA deemed the criterion that innovations possess a technical nature to be a significant factor in the Broccoli and Tomato decisions. Examining the historical documents that preceded the creation of the EPC in 1960, the EBA noted that the legislative architects of the EPC were concerned with excluding from patentability the types of plant breeding processes that were the standard methods for breeding plant varieties at the time. This was because new plant varieties were being created for which a special property right was going to be introduced under the subsequent UPOV Convention in 1960. The sexual crossing of plants judged appropriate for the intended use and the subsequent selection of the plants possessing the required trait or traits were among these traditional techniques. These procedures were distinguished by the fact that the underlying natural phenomena of meiosis determined the characteristics of the plants that resulted from the crossing. The genetic composition of the plants produced was dictated by this phenomena, and the breeder selected plants with the desired trait or traits to get the desired breeding result. The fact that techniques like irradiation that alter plant DNA were mentioned as instances of patentable technical processes further supported the idea that they were to be excluded. The EBA also cited the justifications provided in the Secretariat of the Committee of Experts' memorandum for approving the substitution of the word "essentially" for "purely" biological, which was purposefully done to reflect the legislative intent that the use of a technical device in a breeding process should not be enough to confer a patentable technical character on the process. The EBA came to the conclusion that if a technical step—whether explicit or implicit—is included in a process that relies on plant sexual crossing and subsequent selection, the claimed invention will still be excluded if the technical step merely completes the steps involved in the breeding process.

6.2.3.1. Trademarks

In the marketing of agricultural products, trademarks are quite important. Generally speaking, trademarks are protected by registration. A symbol must be able to be represented in a visible form in order to be registered as a trademark. Names, newly created or preexisting words, characters, numbers, images, symbols, and combinations of these signs are examples of visible signs. A sign must be able to differentiate the products or services of one enterprise from those of other enterprises in order to be eligible for registration. In the majority of countries, marks that are not distinctive, confusingly similar to already-existing marks, or that violate morals or public order are exempt from registration. It has been decided that trademarks that are registered designations of plant kinds are not eligible for protection due to the criterion of distinctiveness. Since AR1 was already a registered plant variety, the attempt to register it as "the name of a registered variety of ryegrass endophyte" was denied. The courts used the test of whether a mark is one that other traders are likely to want to use on or in connection with their goods in the regular course of their business and without any improper motive. When a mark is registered, dealers are prevented from copying it by using marks that are identical or strikingly similar. The majority of registration systems allow for licensing or assignment. To keep track of trademark licenses, a registered user system might be offered. A trademark owner may pursue remedies such as an injunction, restitution orders, and the seizure of products that violate their trademark in the event that a registered mark is violated.

6.2.3.2. Collective and Certification Marks

A collective mark is a unique kind of registered trademark that can be registered by an association whose members are permitted to use it as long as they meet the guidelines outlined in the collective mark usage laws. Therefore, the collective mark's purpose is to educate the public about specific attributes of the product for which it is used. A business that has the right to use the collective mark may also utilise its own trademark. Agricultural cooperatives of produce dealers in the United States employ collective marks. An organisation that promotes the goods and services of its members rather than selling its own products or services is known as the collective mark owner. Use of a certification mark is restricted to compliance with established standards. The primary distinction between certification and collective marks is that the former can only be used by specific businesses, such as members of the association that owns the mark, whilst the latter can be used by anyone who meets the established requirements. In order to support agricultural producers, US state governments often support the registration of certification marks. For instance, the State of Georgia's Department of Agriculture owns the certification mark VIDALIA, which is "intended to be used by persons authorized by certifier, and," according to the Georgia Vidalia Onion Act of 1986, "is used in connection with yellow Granex type onions that are grown by authorized growers within the Vidalia onion production area in Georgia."42 Additionally, non-US agricultural producers have registered certification marks in the USA. FLORIDA CITRUS, which is owned by the State of Florida's Department of Citrus, certifies that the goods bearing the mark "either consist of citrus fruit grown in the State of Florida, under specified standards, or are processed or manufactured wholly from such citrus fruit." For instance, in accordance with the guidelines established by the Thai Ministry of Commerce in "Regulations of the Department of Foreign Trade Re: Usage of the Certification Mark of Thai Hom Mali Rice," Thai Hom Mali rice is recorded as "harvested in Thailand." To certify "that the tea contains at least 100% tea originating in the Darjeeling region of India and that the blend meets other specifications established by the certifier," the Tea Board of India has also registered DARJEELING. It is preferred in Europe that these marks be registered as geographical indicators.

6.2.3.3. Geographical Indications

When a particular quality, reputation, or other attribute of a commodity can be primarily attributed to its place of origin, the indication is known as a Geographical Indication (GI). Article 22.2 of the World Trade Organization's (WTO) Agreement on Trade Related Aspects of Intellectual Property Rights (TRIPS) mandates that nations protect geographical indications. Producers of agricultural products benefit greatly from GIs because they enable them to distinguish their goods from general commodity goods like rice, coffee, and tea, which improves market access. Over 40% of the total earnings from merchandise exports in over 40 LDCs and developing nations comes from the export of a single agricultural item. For instance, more than 80% of Benin's revenue from merchandise exports comes from cotton. More over 70% of Ethiopia's agricultural exports are coffee-related.

Because certain consumers place a higher value on products that are unique based on their origin, several researchers have found that GIs have the ability to command premium prices. For instance, a case study of Comté cheese in France by Gerz and Dupont (2006) revealed that French farmers receive an average of 14% more for milk intended for Comté, and dairy farms in the Comté area since 1990 are 32% more profitable than similar farms outside the Comté area. Babcock (2003) reported that the commodity price for poultry meat in Bresse, France, was quadrupled.

The price of "Zhangqiu Scallion" per kilogramme increased from 0.2–0.6 yuan before to the adoption of the origin mark to 1.2–5 yuan in 2009, accordin□ to Kireeva, Xiaobin□, and Yumin's (2009) analysis of the ori□in mark's use in the PRC. The price of "Jianlian" lotus seed increased from 26–28 yuan per kilogramme to 32–34 yuan per kilogramme when it was registered as a GI in 2006. Traditional crops are frequently freer of contaminants like pesticides and herbicides, and GIs applied to these products give consumers confidence in their safety and traceability, which is one of the reasons why GI-marked products command higher prices (Blakeney, 2017). Producer associations from the appropriate region typically keep an eye on and safeguard geographic indications. Because a user must have the proper relationship to the geographical location and adhere to the production processes of that region, geographical indicators, unlike trademarks, are not freely transferable from one owner to another. By registering, geographic indications are acquired. Typically, a specification is filed that lists the pertinent □eo□raphic re□ion and the attributes of product quality that are specific to that re□ion. Typically, a body that represents the local producers files the re□istration application. Typically, this authority is also in charge of pursuing legal action against improper users of the geographical indicator.

6.2.3.4. Trade Secrets

The law of confidence may safeguard information, including know-how, that has been created by an individual, is not in the public domain, and about which efforts have been taken to maintain confidentiality. Theft of genetic material has been ruled to be actionable. For instance, in Franklin v. Giddins46, the Queensland Supreme Court ruled that the defendant had stolen confidential information contained in the budwood's genetic composition. The defendant had stolen budwood cuttings from the plaintiffs' orchard, which allowed the defendant to grow early-ripening nectarines after grafting.

The US Ei□hth Circuit Court of Appeals addressed a conflict between rival maize seed breeders in Pioneer Hi-Bred International v. Holden Found Seeds47. Holden had created a seed using stolen seed, which Pioneer said was one of its trade secrets. Holden denied that its seed and Pioneer's H3H/H43SZ7 shared any genetic similarities. The court managed three sets of tests—electrophoresis, reverse phase high-performance liquid chromatography, and □row-outs—in an effort to assess the parties' conflictin□assertions. The parties kept an eye on each test, which was conducted by impartial experts under the court's supervision. Pioneer was awarded $US46 million for the theft of its trade secrets after the court determined that Holden's explanation of the seeds' paternity was implausible and preferred Pioneer's interpretation.

6.2.3.5. Copyright

The preservation and exploitation of the tangible manifestation of ideas are the focus of copyright law. This can include a computer program or collection of computer programs, as well as a table or compilation that is conveyed in words, figures, or symbols. Therefore, scientific publications, scientific databases, lab notebooks, and computer displays of data may all be protected by copyright. The main way that copyright law relates to agricultural research is by implying that, in addition to or instead of requesting a patent or other kind of intellectual property protection, copyright may be claimed over textual representations of genes or amino acid sequences.

6.2.3.6. Biodiversity Laws

The international biodiversity regime is becoming more and more important for agricultural research organisations to consider. The Nagoya Protocol, 201048, the International Treaty on Plant Genetic Resources for Food and Agriculture, and the 1992 United Nations Convention on Biological Diversity (CBD) all aim to create an international framework for the preservation and use of the world's biological resources as well as for the "fair and equitable sharing" of the advantages that result from their use. Indigenous peoples' traditional knowledge has been crucial in determining which biological resources are suitable for economic exploitation worldwide (Blakeney, 2002). The application of traditional farmers' and indigenous peoples' expertise has also made a substantial contribution to biodiversity conservation and the creation of new crop varieties. These organisations have played a significant role in the preservation of plant genetic resources and their dissemination to seed firms, research institutions, and plant breeders. Benefits from the use of those resources must be shared with farmers and traditional people in a number of nations, including India.

Chapter 7

Food Security and Intellectual Property Rights

Contents

OVERVIEW

Hunger is a grave violation of human rights and dignity. It hinders development fundamentally, increases crime and violence, lowers productivity, and shortens life expectancy. More than 800 million people, mostly in developing nations, lack access to enough food to meet their basic nutritional needs, according to a 1996 World Food Summit organised by the Food and Agriculture Organisation of the United Nations (WFS) in Rome. Malnutrition was thought to kill almost 400,000 individuals per day. At the Rome Summit, the 185 participating nations pledged to provide universal food security. Food security is the ability of all peoples to always have physical and financial access to enough wholesome, safe, and nourishing food to satisfy their dietary needs and food choices and lead active, healthy lives. The Summit's Rome Declaration promised to reduce the number of hungry people by half by 2015.The United Nations' Millennium Declaration from 2000 also contained this objective. In order to achieve this goal, the number of undernourished people has to decrease by 20 million year. Data from 2001, however, showed that the annual reduction was less than 8 million. According to estimates, the world's demand for cereals will rise by 20% between 1995 and 2020 at the current rate of population growth, and developing nations' net imports of cereals will need to double in order to keep up with demand. The developing world already imports 88 million tonnes of grains annually at a cost of $14.5 billion, and between 1995 and 2020, the world's cereal demand is expected to rise by 40%. Ironically, a 1999 research by the International Food Policy Research Institute (IFPRI) predicted that, at least until 2020, the world's food supply would continue to expand faster than population growth. Food insecurity in emerging nations can be attributed to a variety of issues. Poverty and limited access to food are two of the main causes of these issues. More than 1.3 billion individuals on the planet make less than $1.00 USD a day, while another 2 billion make only slightly more. In order to achieve food security, poverty must be eradicated, according to the Plan of Action approved at the WFS. The significance of having access to land, water, better seeds and plants, suitable technologies, and farm credits was emphasized. Increases in cereal yield, the area of arable land, and the use of fertilizer and insecticides all contributed to the enormous increases in food productivity that occurred during the Green Revolution, which lasted for 30 years from 1960–1990. Increasing agricultural production on the currently available land while preserving the base of natural resources is necessary to satisfy the food security demands of the next 30 years and to generate wealth in underprivileged areas. Economically and environmentally, it is no longer acceptable to rely on the high-yielding, chemically cultivated crop varieties of the past. Modern agricultural science and traditional agronomic knowledge must be combined in a second

Green Revolution. Intellectual property rights are a key component in this field of agricultural innovation.

7.1. ISSUES CONCERNING IPR AND FOOD SECURITY

7.1.1. PRIVATIZATION OF GERMPLASM

It has been argued that private intellectual property rights should not be used to sterilize food security because it is too vital. The concentration of proprietary biotechnologies in a small number of firms as a result of the prophetization of genetic resources is a related worry.99 There are "six major industrial groups who between them control most of the technology which gives the freedom to undertake commercial R&D in the area of GM crops," according to the Nuffield Council's report on bioethics and genetically modified crops. They are:

- Agrevo/Plant Genetic Systems;
- ELM/DNAP/Asgrow/Seminis;
- Du Pont/Pioneer;
- Monsanto/Calgene/Delkalb/ Agracetus/PBI/ Hybritech/Delta and Pine Lane Co.;
- Novartis; and
- Zeneca/Mogen/Avanta.

The British House of Lords Select Committee, in its 1998 report on the EC Re□ulation of Genetic Modification in A□riculture, also cautioned about the issue of cartels and monopolies in the agrochemical/seed industry, pointing out that the level of consolidation was already si□nificantly hi□her than in the pharmaceutical industry. About 60% of the 410 patents pertaining to the Bt gene and Bt pesticide technology were held by the six largest industry groups at the time, according to a 1997 study by Krattinger on the evolution of insect resistance in crops. Research on the alteration of cry proteins, which have a selective application to the different agricultural pests, was sealed off as a result of this concentration of patent ownership. One crucial subject that requires empirical research is how oligopoly tendencies in the biotechnology market affect international organizations' ability to supply public goods to developing nations in the agriculture sector. In his historical analysis of pharmaceutical patentin□, Dutfield (2017) documents the cartelized use of patentin□as a means of market protection and competition. Modern "life sciences" businesses have a patent legacy and culture that dates, in some cases, back to the turn of the century because they were primarily separated from the pharmaceutical patenting industry. The market dominance of these private firms has a si□nificant impact on the type of biotechnological research that is conducted, in addition to the potential negative effects that this market concentration may have on the strength of competition. How far, for instance, would the predominance of private companies in agricultural and biomedical research steer it towards Northern issues, such as away from Southern food priority and health issues? Will public research institutions be able to get IPRs in critical enabling technologies on reasonable terms from their owners? The "biomedical anti-commons tra□edy," as defined by Heller and Eisenber□106, is the result of patent office's □rantin□ overly broad patent claims, which has further concentrated proprietary technologies in the hands of a relatively small number of Northern life-sciences companies. As previously stated, policy directives to patent offices can only address this issue to a certain degree because, in the end, the courts have the last say over how to interpret patent claims.

The TRIPS Agreement's Article 27(3)(b) allows WTO members to exclude plants, animals, and fundamentally biological processes used in the production of plants and animals from patent protection. In particular, members are not allowed to exclude microorganisms, non-biological processes, or microbiological processes from patent protection. The wording of Article 27(3)(b) suggests that microorganisms and plants and animals can be clearly distinguished from one another. However, neither in science nor in patent office practice is there a widely reco□nized definition of "micro-or□anism." The absence of a definition allows members to restrict this exclusion from patentability in very different ways. For instance, it has been noted from the patent activity occurring in the US, Europe, and Japan that the idea of patentable subject matter is interpreted quite loosely, with an emphasis on inclusion rather than exclusion. It appears that there is no apparent need for a definition based on the practices of patent granting agencies in developed nations. The main concern for protection is not the invention's subject matter but rather whether it satisfies the requirements for patent issuance. The European Patent Office (EPO) cited "it does not seem expedient to introduce such a definition as the rapid evolution in the field of microbiology would necessitate its frequent updating" as one of the reasons for the reluctance to employ a definition. Because the scientific classification is always chan□in□, the word "micro-or□anism" is not well defined in scientific practice. The only discussion of any kind in European patent office practice has concerned the distinction between a plant, which is patentable, and a plant variety, which is not. As a result, patent □rantin□ offices have a□reed that excluded material is that which is protected by a plant variety right as established by the UPOV Convention; anything that does not fall under the purview of UPOV is patentable. As a result, groups of plants that include multiple varieties can be protected by patents as lon□ as the patent holder hasn't specifically claimed a particular variety. Therefore, everything that is not a patent variant in the traditional sense is patentable. No more definitions have been thou□ht to be required. The awarding criteria are interpreted broadly, and the patent laws of developed nations are based on the presumption of patentability. Therefore, any exclusions are subject to a limitation application. As a result, when a nation adopts specified cate□ories of prohibited content, these exclusions are likely to be closely examined, especially if the categories may be deemed to exceed what is allowed by the TRIPs Agreement.

It could be wiser for member states to establish a higher bar for patent protection for livin□ material □iven the challen□es involved in tryin□ to define microorganisms. Adcock and Llewelyn, for instance, argue that an innovation containing biological components might not qualify as novel:

- If the information is already in the public domain; and/or
- The invention merely replicates biological material, or the function of biological material, which already occurs naturally.

An invention involving biological material will be regarded as lacking an inventive step if it:

- Merely identifies the biolo□ical material; and/or
- Merely identifies the natural function of the biolo□ical material.

If an invention involves a major technical application of a recognized biological material function, it will be considered an innovative step. This technical application must be a substantial technological advancement over the previous art and must go beyond a straightforward copy of the biological material's natural function. If it can be demonstrated that an invention containing biological material can be employed in a way that offers a measurable public benefit, it will be considered capable of industrial application. For an invention to be considered public, it must be able to be applied in a way that promotes social, environmental, and economic well-being as well as public health. As evidenced by the revocation of the Neem and Turmeric patents, the courts are increasingly deciding what constitutes patentability due to the low protection standards currently enforced by the US and European patent offices. The bioscience industry will benefit from increased predictability and certainty as a result of raising the threshold for protection, which will guarantee that inventions that merit protection are protected and that this protection is less likely to be later contested in court. Examples of judges being compelled to reevaluate the patent offices' liberality include the reopenin□ of the Neem and Turmeric patents. However, these could be viewed as illustrations of why patent offices need access to traditional knowledge data in order to maintain the state of the art. Regarding the patenting of biotechnological inventions, general guidelines cannot be made for all emerging nations. "At this time, there seems to be little evidence that providing patent protection for biotechnology-related inventions is really in the interests of the majority of developing countries which have little or no capability in this technology," the UK Intellectual Property Commission concluded in its Final Report on Integrating Intellectual Property Rights and Development Policy. For this reason, we would advise making the most of TRIPS' ability to bar such ideas from patent protection. On the other hand, nations that now have or hope to create biotechnology-related sectors may offer some forms of patent protection, with particular exclusions for research and plant breedin□ as well as a definite exemption for farmers who want to reuse seeds.

7.1.2. IP and Agricultural Research

The public agricultural research system's ability to perform its public good purpose of helping to eradicate food insecurity is called into question by the commercialization of genetic resources and enabling technology. "It is likely that much research in biotechnology and agriculture will end up as an international rather than public good and that it will be distributed according to complex licensing structures," mentioned Drahos. In order to better address the nutritional needs of rural communities, many farmers with limited resources grow minor food crops rather than just big crops like maize, rice, and wheat. For instance, villages may cultivate over 150 crop species and cultivated variations throughout Nepal's hills and valleys. However, in general, breeding for minor crops with tiny markets is not encouraged by plant variety protection (PVP). This is due to the fact that breeders will receive very little return on their investment in research. Instead, they promote breeding aimed at important crops with high commercial potential. Furthermore, certain protected plant varieties could not even qualify as food crops. For instance, until relatively recently, foreign-bred roses grown for sale made up over half of Kenya's protected new kinds. Therefore, it is possible that PVP could be a factor in the trend of monocultures of single agrochemical-dependent crop varieties replacing traditional diverse agro-ecosystems with a wide variety of traditional crop varieties. This would reduce the variety of nutrient-dense foods available in local markets. This trend, which began before PVP systems were introduced, is undoubtedly a worldwide phenomena. Nonetheless, it is one that PVP's existence and growing popularity may subtly promote. However, there is nothing stopping developing nations from promoting research on these small crops that are significant to their communities, either by strengthening the intellectual property rights of these species or by implementing non-IPR-related policies like public subsidies.

There is a worry that high-value crops like flowers may replace food security crops when PVP systems are implemented in underdeveloped nations. Food security may stabilize as a result of the ensuing change in research objectives, which may provide challenges for technological transfer and development aimed at addressing food scarcity issues. The amount of land available for the production of native crops with lower export values would decrease, even as the food and agricultural industries' export performance is anticipated to improve. The security of food supplies may suffer as a result. However, the system might still be advantageous overall, even for the poor, if the proceeds from the sale of higher-value products go to them.

7.1.3. IP Protection of Plants and Seeds in Developing Countries

The impact of PVP on nutrition and food security in developing nations is discussed, which makes one think more broadly about the suitability of such an IPR for these nations. Sadly, there aren't many empirical studies available to us. One of the few looked at "the (expected) impact of plant breeders' rights (PBR) on developing countries with respect to: private investment in plant breeding, breeding policies of public institutes, transfer of foreign germplasm, and diffusion of seed among farmers." It was a 1994 joint project of the Anti-American Institute for Cooperation in Agriculture and the University of Amsterdam. Five nations were utilised as case studies; two (Colombia and Mexico) were set to implement PVP systems, while three (Argentina, Chile, and Uruguay) already had them in place. These nations are comparable in that they essentially have two seed markets. While domestic companies dominate the seed industry for self-pollinating varieties, multinational organisations control the hybrid seed market. Argentina stands out from the others, though, as the only nation whose PVP right holders have effectively upheld their rights to the point where they have a same amount of control over the supply of wheat and soy seeds as their American counterparts. As a result, the study's authors came to the conclusion that PVP in that nation has most likely "triggered the reactivation of some soya bean breeding programs and prevented the local wheat companies from reducing or even terminating their breeding activities." There isn't much proof that PVP has made it much easier for domestic seed businesses to obtain current cultivars, unique genetic stocks, and foreign genomic material when it comes to exotic germplasm. Additionally, businesses that have permits to use patented varieties from foreign breeders may occasionally face limitations on the countries to which they can export. For instance, the United States breeder and the European licensees did not want the plantlets to compete with those already produced in Europe, therefore in 1994, Argentine strawberry plant farmers were not allowed to export their plantlets to Europe. Public agricultural research centers in Chile and Argentina use PVP to seek funding and work with businesses. The chapter claims that this is changing the focus of public research and making their germplasm less accessible to the general population. What impact do farmers face? First, seed merchants in Argentina are now required to pay taxes and royalties on the seeds they sell. These expenses have not yet been transferred to the farmers. Second, transplanting farm-saved seed has not been hindered by PVP laws in the three nations where they are well-established. Third, the chapter states that PBR primarily favors plant breeding for farmers who operate in reasonably prosperous conditions because many current plant types are inappropriate for resource-poor farmers.

Another intriguing example is Kenya. With the passage of the Seeds and Plant Varieties Act, which went into effect in 1975, Kenya became one of the first developing nations to adopt PVP legislation. The lack of support for such legislation from domestic breeders indicates that legislators were mostly convinced of the necessity of such a law by outside pressure, primarily from foreign companies whose horticultural varieties were being planted in the nation. Protected varieties must be sufficiently distinct, sufficiently varietal pure, sufficiently uniform or homogenous, and stable in their basic traits, according to the Act, which is mostly based on the UPOV Convention and the corresponding UK legislation. It is also necessary for "the variety's agroecological value to surpass, in one or more characteristics, that of existing varieties according to results obtained in official tests." However, it wasn't until the 1990s that the Seeds and Plant Varieties (Plant Breeders' Rights) Regulations were passed in 1994 and the Plant Breeders' Rights Office was founded in March 1997 that the PVP section of the act could be put into effect. The Registrar of the Plant Breeders' Rights Office claims that during the first few years, the majority of the more than 200 applications—116 of which were from foreigners—were for horticultural varieties, with roses accounting for over half of the total. Only recently has the public sector, which creates the majority of newly bred varieties in Kenya, expressed interest in applying for protection. These days, its uses are growing. Even while new businesses are opening, it will probably be a few more years before any increase in private sector breeding activities is mirrored in an increase in applications because of how long it takes to breed new types. It is probably too soon to determine if the system has a beneficial or negative impact on food security or how likely it is that other developing nations would follow Kenya's example. The PVP system's greatest beneficial function at the moment is most likely to promote the importation of foreign-bred cultivars into Kenya. This is required for items that are grown primarily for export and that rely significantly on foreign breeding material. Cut flowers are possibly the most significant of these. Other developing nations lack experience in protecting plant varieties. There aren't many more models besides UPOV. The OAU, Heads of States Summit in Ouagadougou in June 1998 adopted the African Model legislation for the Protection of the Rights of Local Communities, Farmers, and Breeders, and for the Regulation of Access to Biological Resources. This legislation adopts a sui generis regime based on UPOV 1991, but incorporates Farmers' Rights and combines them with some of the CBD's access principles. India's Protection of Plant Varieties and Farmers' Rights Act, 2001, is probably one of the most significant PVP legislation in developing nations. The Act's principal goals are:

- To encourage governmental and private sector investments in research and development for the creation of novel plant types by guaranteeing suitable returns on such expenditures;
- To support the expansion of the nation's seed business through both foreign and indigenous investment, guaranteeing Indian farmers access to high-quality seeds and planting materials; and
- To maintain farmers' traditional rights and compensate traditional, rural, and tribal communities for their contributions to the nation's agrobiodiversity, while also acknowledging the role of farmers as cultivators and conservers.

Although there are some parallels to UPOV 1978, more clauses are added to safeguard the interests of farmers and public sector breeding facilities. The Act, for instance, protects "a farmer's right to save, use, exchange, share, or sell his farm produce of a protected variety," with the exception of "in case the sale is for the purpose of reproduction under a commercial marketing arrangement." Applications for PVP "must be accompanied by an affidavit sworn by the applicant that such variety does not contain any gene or gene sequence involving terminator technology," according to another intriguing clause. From private sector seed firms to public enterprises and research institutions, as well as resource-poor farmers, the Act seems to represent a sincere endeavour to apply TRIPS in a way that respects the unique socioeconomic interests of all the different producer groups in India. The 2001 legislation will undoubtedly need to be changed, though, as India is already joining UPOV under its 1978 Act. The extent to which germplasm and other genetic resources from developing nations are patented in industrialised nations and the amount to which this material is patented in developing nations is not well documented. 400 patents that have been cited to the USPTO and EPO are found in our sample research on the patenting of genetic resources from the Asian region. This is a serious misuse of resources in developing nations, and it is recommended that procedures be put in place immediately to get source countries' informed consent and appropriate benefit-sharing plans. Comparable data regarding the level of germplasm patenting in emerging nations is currently lacking. Since it is doubtful that developing nations will be the markets for products that use the patented material, it would be unrealistic to expect this to occur. "Defensive patenting" is an intriguing application of the patenting system to maintain genetic material in the public domain. In nations that use a global test of novelty, patenting genetic material—or even merely filin□provisional patent applications—will have the impact of making the information about that material publicly available. However, there will be an issue with the USA, which has a national novelty test.

7.1.4. Plant Variety IPR and Poor Farmers Interests

Farmers in developing nations, where a sizable section of the population depends on agriculture for job and money, are particularly concerned about the issue of seed preservation. Since many of these farmers are small-holders, they frequently trade seeds over the fence and save harvest seeds for future plantings. This is particularly true in nations where seed production and distribution are not significantly influenced by the public or private sectors. Despite permitting on-farm replanting, the 1991 amendment to the UPOV system's regulations limits farmers' ability to purchase seed from sources other than the original breeders or their licensees. Only if the seed is from their own harvest and will be used for replanting on their own farm—and not a fruit, tree, or ornamental species—do the following nations allow farmers to save seed:

- **Africa & Middle East:** Equatorial Guinea, Gabon, Guinea, Guinea-Bissau, Mali, Mauritania, Morocco, Niger, Senegal, Togo, Benin, Burkina Faso, Cameroon, Chad, Central African Republic, Congo, Cote D'ivoire, and Egypt. Only if the original seed was obtained lawfully (that is, from a source approved by the breeder) and replanted on the farmer's own property (no sharing, trade, or selling) can it be conserved in the South African case.
- Asia:
  - **China:** Only if the seed is from their own harvest and is replanted on their own farm.
  - **India:** Only without packaging and selling the farm-saved seed as a protected variety
  - **Republic of Korea:** Only if the seed is from their own harvest and is replanted on their own farm, only up the amount of seed needed to replant own farm
  - **Philippines:** Only for and among small farmers, any exchange or sale must be for reproduction or replanting on own land, selling the variety under the trade mark or name associated

with it is expressly prohibited and subject to further conditions and guidelines to be issued

- **Thailand:** Only if the seed is from their own harvest and for replanting on their own farm and if the variety is published as a "promoted" plant variety, then its "cultivation or propagation by a farmer may be made in the quantity not exceeding three times the quantity obtained"
- **Vietnam:** Provided the farm-saved seed is not sold.

• Latin America & Caribbean:

- **Argentina:** Only if the seed is from their own harvest for replanting on their own farm and is neither a fruit species nor an ornamental plant.
- **Bolivia:** Only if it is for their own use and if it is not a fruit, ornamental or forest species and only for farmers with a landholding of 200 ha or less under cultivation, of which a maximum allowance per crop within that area is as follows: 100 ha for soybean, wheat, maize, sor□hum, sunflower or cotton; 50 ha for rice; 20 ha for all other crops

  Brazil: Only small farmers, and no exchange of seeds except if it's on a non-commercial basis with other small farmers and not if it's sugar cane
- **Chile:** Only if the variety was legally acquired (i.e., a royalty was paid) and only to replant it on their own farm.
- **Colombia, Ecuador, Peru and Venezuela:** Only if it is for their own use and if it is not a fruit, ornamental or forest species
- **Nicaragua:** Only if it is for their own use
- **Panama:** Only if the variety was legally acquired to be replanted on their own farm
- **Trinidad & Tobago:** Not for commercial marketing or sale
- **Uruguay:** Only if the seed was replanted on the farmer's property, was legally obtained (i.e., royalties were paid to the breeder), and was maintained apart from other seeds throughout the entire process (harvest, cleaning, conditioning, etc.). Any amount of seed kept on the farm that seems excessively big considering the size of the farm is the grounds for a supposed legal violation, and the free interchange and commercialization of farm-saved seed are specifically forbidden.

Concerns about food security have not yet been included by any nation in the implementation of plant variety ri□hts protection. Seed firms contend that farmers are free to continue growing non-plant variety-protected seed, including traditional local varieties, if they so want, negating the need for them to buy PVP-protected seed. Furthermore, government assistance for farmers, includin□ loans, is frequently contin□ent on the plantin□ of specific crops and seed varieties, such hybrids, in many developing nations. Additionally, providers frequently employ seed aid to encoura□e the usa□e of specific crops and seeds. It should be mentioned that farmers may not always find preservin□ seeds to be a financially advanta□eous choice. "The cost advanta□e of savin□ seed is eroded by the deterioration of saved seed, causing yield losses over time," according to van Wijk et al. Even though hybrids deteriorate particularly quickly, some Latin American farmers are choosing to save hybrid maize for a second □eneration due to the si□nificant discrepancy between the cost of savin□ seeds and the price of new seeds. However, it should be remembered that buying seed is just one of many agricultural inputs that need to be paid for, and even low-income farmers may choose to spend more for higher-quality seed if they anticipate a larger crop. The "sui generis" clause in TRIPS does grant nations some latitude to modify national PVP systems to address such issues, regardless of the precise truth. As a result, some developing nations are creating alternative PVP systems that partially aim to improve food security, even as more join UPOV. They accomplish this, for instance, by permitting farmers to obtain protected seed from any source and/or mandating that protected cultivars have traits that are actually better than those of current cultivars. The African Union (formerly the Organisation of African Unity) has created a model law for member governments to consider, known as the African Model Legislation for the Protection of the Rights of Local Communities, Farmers and Breeders, and for the Regulation of Access to Biological Resources. The Indian parliament has also passed legislation that would preserve farmers' freedom to save, sell, and exchange all produce of a protected variety. The interests of farmers and breeders are given equal weight in both situations.

7.1.5. Plant Variety IPR and Genetic Resources Availability for Breeding

The need for unrestricted access to genetic material, particularly that which is protected by intellectual property rights, has been emphasized by plant breeders. For this reason, a broad breeders' exemption is included in the UPOV Convention. The research exemption under patent law is typically far more restricted to noncommercial scientific or experimental use. The recent US decision in Madey v. Duke University120, which ruled that a university that entered into commercial research contracts could not use the defense, exemplifies the narrowness of the research exception in patent law. In Monsanto v. Stauffer, the scope of the experimental research exception in UK patent law was investigated. In that instance, Stauffer had created a commercial version of Monsanto's popular patented weed killer "Roundup," called "Touchdown," for which they had received temporary approval from the appropriate authorities. Stauffer had set up tests at their own research farm and arranged a number of tests outside the farm so that interested parties could see the results in order to secure final permissions. The patent court approved Monsanto's request for an interlocutory injunction on the basis of patent infringement, rejecting the company's argument that testing their product in various soil and climate conditions outside of the research farm amounted to an experimental use. The Court of Appeal excused all studies conducted at Stauffer's research farm as well as in UK labs and greenhouses, but acknowledging that tests conducted outdoors could not be eligible for an experimental use exception. According to its size, scope, target audience, and technique, the Court restricted how the term "experimental" may be interpreted. The extent to which university researchers could utilise the experimental use exception to field trials has been called into question by this case. Another factor contributing to the experimental use exception's relative narrowness in patent law when compared to laws protecting plant variety rights is that, although a protected plant variety is covered by a single title, biotechnological inventions related to plants are more likely to be protected by a patent, and occasionally multiple patents. In addition to plants, seeds, genes, and DNA sequences may also be covered by the patents. Patents have the consequence of limiting access to the patented "products." Patents have been criticized for "locking up" genetic resources because plant breeding innovation is cumulative and relies on having access to a large supply of material. The FAO International Treaty introduced several clauses, as outlined above, to address this concern. IPRs may not be the only reason for the limitations on access to breeding material, though. For starters, certain nations have decided to exclude from the multilateral system to be established under the International Treaty specific types of plant genetic resources that they believe to be strategically significant. Additionally, some poor nations have limited the free flow of their genetic resources by using their CBD rights to control access to them. Even in their own nations, this might be harmful to long-term food security. The link between IPRs and the privatization of agricultural research, the decline of nonproprietarial public sector research, and the growing concentration of ownership of breeding material, research tools, and technologies in the hands of a few large corporations goes beyond these concerns about how particular intellectual property rights privatize genetic material required for breeding.124 This trend can hinder the implementation of public policies intended to improve food security in addition to limiting the free flow of breeding material. This is due to the fact that governments have a far harder time influencing businesses than the public organisations they partially or fully support.

7.1.6. Farmers Rights and Traditional Knowledge

Indigenous peoples' and traditional farmers' expertise has been crucial to the creation of new crop varieties and the preservation of biodiversity. These organisations have played a significant role in the preservation of plant genetic resources and their dissemination to seed firms, plant breeders, and research organisations. While breeders and seed corporations have used intellectual property rights to recoup their development costs, they have generally not been compensated for the value they have provided. Farmers that use enhanced varieties, however, are required to pay for them. It is challenging to estimate the economic value of the biological diversity that traditional farmers preserve for agriculture. Because "conventional breeding increasingly focusses on crosses among elite materials from the breeders' own collections and advanced lines developed in public institutions," it has recently been proposed that "the value of farmers' varieties is not directly dependent on their current use in conventional breeding, since the gene flow from landraces to privately marketed cultivars of major crops is very modest." However, the germplasm supplied by traditional tribes is frequently the source of those collections and advanced breeding lines. The degree to which biodiversity offers a pool of domesticable species and genetic resources for improving domestic species is becoming an increasingly important economic asset. The current biotechnology revolution has enabled the engineering of desired genetic traits from beneficial indigenous species. Approximately 6.5% of all agricultural genetic research focuses on germplasm from wild species and land races. When it comes to creating farming methods and practices that are tailored to the local environment, traditional knowledge is especially crucial. In addition to helping to manage the environment by reducing erosion, preserving soil fertility, and promoting agrobiodiversity, this may make it possible to use marginal lands, which would increase food security by facilitating access to food in remote locations.

The so-called Basmati controversy serves as an example of how traditional knowledge, farmers' rights, and food security are intertwined. This started when the USPTO awarded a patent for "Basmati rice lines and grains" to RiceTec, an American business with headquarters in Alvin, Texas, in September 1997. Similar to the Neem debate, this was seen as yet another instance of alleged "biopiracy" of Indian genetic resources. According to the patent, these "novel rice lines" "produce rice grains having characteristics similar or superior to those of good quality basmati rice" and "have plants that are semi-dwarf in stature, substantially photoperiod insensitive, and high yielding." At least two of the 15 known semi-dwarf long grain types and any of the 22 mentioned basmati known lines were involved in the crossing; "any selection and propagation method or scheme known in the art may be used." The Research Foundation for Science, Technology, and Ecology, an NGO, petitioned the Indian Supreme Court in March 1998 to order the government to either contest the contentious patent or file a complaint with the WTO's Dispute Settlement Body. Finally, alle□edly in April 2000, the Indian □overnment filed a complaint with the USPTO contestin□ three claims—claims 15–17—that included a broad description of the rice grains in question. In August 2001, the USPTO revoked claims 15–17 and several others that were not being challenged either (especially those pertaining to the process). This left only the claims pertaining to the rice lines that RiceTec had effectively bred. As a result, RiceTec withdrew four claims, one of which was not being challenged. In parallel, the US Federal Trade Commission received a petition from the Research Foundation for Science, Technology, and Ecology and two American non-governmental organisations based in Washington. The petition was based on the FTC's implementing regulations and the First Amendment's right to petition the government. "Where a geographical origin has acquired a peculiar trade si□nificance, the use of a term descriptive of such article cannot be applied to a product of a different origin, even though such a product may be of identical quality ,"" the petition asserted, citing an FTC cease and desist order from a prior case (the "New River" case). However, the FTC emphasized in its May 9, 2001, response that its "efforts are focused on those areas which may affect the greatest number of consumers, may pose a risk to consumers' health or safety, or may cause si□nificant economic harm to consumers… we do not have reason to believe that si□nificant consumer injury is likely to arise from current rice marketing." Basmati and jasmine rice are examples of "aromatic rough rice" under US Department of Agriculture rules, and they are not restricted to rice cultivated in any one nation. Therefore, using terms like "basmati" or "jasmine" to describe rice farmed in the United States is not specifically restricted by law or re□ulation. The Biolo□ical Diversity Act, which was passed on December 11, 2002, and the Protection of Plant Varieties and Farmers' Rights Act 53 of October 30, 2001, were two measures that the Indian government had developed in the meantime to implement the Convention on Biodiversity (CBD). However, there would have been no way to declare the patent void on the grounds that it was foreseen by an Indian plant variety rights (PVR) certificate as lon□as the rice lines mentioned in RiceTec's invention were not protected prior to the application. This would have been even less likely because, according to US patent law, a foreign previous title or publication must have been published at least a year before the USPTO application was submitted in order to qualify as an anticipation. Therefore, these laws are intended to bring

Indian laws into accordance with international treaties that India has accepted and to lessen the necessity for future proceedings to invalidate foreign patents.

"The responsibility for realizing Farmers' Rights, as they relate to Plant Genetic Resources for Food and Agriculture, rests with national governments," according to Article 9.2 of the FAO International Treaty on PGRFA. National legislation should include provisions pertaining to "(a) protection of traditional knowledge relevant to plant genetic resources for food and agriculture." For subsistence people, traditional wisdom has been crucial in ensuring food security. In order to ensure their yearly harvests, a minimum level of production, and to avoid food shortages, farmers in subsistence systems have tended to use a wide variety of crop species. In many cases, seed production has relied on gathering and domesticating wild, regionally recognized types. Crop species that support productivity and disease resistance are essential to modern agricultural techniques, and they can only be sustained with the constant addition of new □ermplasm. Formal breedin□ techniques benefit □reatly from the diversity of landraces and the information that is available about their unique characteristics. It has been observed that the decline in traditional knowledge coincides with the decline in biological variety. When a plant variety goes extinct, all of the information regarding its characteristics is doomed to become irrelevant. "The responsibility for realizing Farmers' Rights, as they relate to Plant Genetic Resources for Food and Agriculture, rests with national governments," according to Article 9.2 of the WTO International Treaty on PGRFA. National law should contain measures relating to:

- Preservation of customary knowledge pertaining to plant genetic resources for agriculture and food;
- The right to fairly share in the advantages that come from using plant genetic resources for agriculture and food; and
- The right to take part in national decision-making regarding issues pertaining to the preservation and sustainable use of plant genetic resources for agriculture and food.

Lastly, Article 9.3 states that it cannot be read "to limit any rights that farmers have to save, use, exchange, and sell farm-saved seed/propagating material." Article 9.1 makes the assumption that the landraces that traditional farmers use constitute a dynamic genetic reservoir for the transmission of desirable genetic features and the generation of new types. Local and indigenous groups' traditional wisdom is seen similarly. There can be little disagreement over the amount and distribution of this compensation, which is intended to honor these groups for their prior contributions to the creation of plant genetic resources for the production of food and agricultural products. Any determination of the fair share that traditional farmers and indigenous communities would receive under a regime based on traditional knowledge or farmers' rights will always be arbitrary. But since arbitrary estimates are nothing new to the intellectual property system, the 20-year patent period is meant to give all innovators, regardless of technology, a chance to be compensated. Likewise, the UPOV Convention's 25-year exclusivity for new tree and vine varieties ignores the differences in R&D expenses among the various types. Plant breeding and plant patenting are the main methods by which plant genetic resources are converted into food and a□ricultural produce. The principle of fair benefit sharin□ with farmers for their contributions to plant breeding and patenting advances forms the foundation of a farmers' rights framework. Germplasm from wild species and land races is thought to be the subject of roughly 6.5% of all genetic research conducted in agriculture. In accordance with their national laws, the Contracting Parties to the Plant Genetic Resources Treaty are required by Article 9.2 "to take measures" to safeguard and advance farmers' rights. The protection of traditional knowled□e, fair benefit sharin□, and the ri□ht to take part in decision-makin□ are all included in the content of these ri□hts, which are defined by the balance of that clause. The legal framework under which Farmers' Rights are to be implemented is left open by the Treaty. One of the UPOV versions and parts of the CBD's access principles are frequently combined in national legislation pertaining to farmers' rights. The OAU Heads of States Summit in Ouagadougou in June 1998 enacted the African Model Law for the Protection of the Rights of Local Communities, Farmers and Breeders, and for the Regulation of Access to Biological Resources, which adopts a sui generis regime based on UPOV 1991. The majority of national laws, however, favor access laws in conjunction with UPOV 1978 (e.g., Kenya's Seeds and Plant Varieties Act 1975; Costa Rica's Biodiversity Law 1998; India's Community Intellectual Property Rights Act 1999; and the Andean Community's Common System on Access to Genetic Resources, 1996).

7.1.7. Conservation of Biological Diversity and Food Security

According to some theories, intellectual property rights (IPRs) offer distorted incentives that promote actions that harm biodiversity and, consequently, harm food security. Three sets of questions are brought up in relation to this:

- Does the expansion of monocultural agriculture become aided by intellectual property rights? And if so, does biodiversity deteriorate as a result?
- Do plant variety rights promote the use of a comparatively small pool of genetic material and the cultivation of genetically homogeneous varieties? And is biodiversity negatively impacted if one of these is true?

- Does biodiversity suffer as a result of the growing manufacturing and distribution of seed-agrochemical "packages" (such as transgenic crops that are marketed with pesticides and/or herbicides to which they are naturally resistant)? If so, do IPRs encourage businesses to create this type of "package"? Is this an IPR issue, to put it another way?

Re□ardin□ the first set of issues, Reid's criticism of IPRs is amon□ the most credible since he finds a direct link between IPRs and a tendency towards centralized research, which he feels affects agro-biodiversity. "Centralized crop breeding and the creation of uniform environmental conditions, and discourages agro-ecological research or local breeding tailored to local conditions," he concludes, referring to the current policy framework for the use of genetic resources for food and agriculture. IPRs increase the incentives for creating seeds with high potential demand. In order to maximize demand for their products, seed companies typically concentrate their research on high-value crops that are commonly used and create cultivable variations. This entails either releasing the new seeds while simultaneously encouraging farming methods that lessen environmental variability or breeding through gene selection for optimal adaptation. This IPR-supported bias in favor of centralized crop breeding programs has the following detrimental effects on biodiversity:

- Decreased crop diversity;
- Decreased spatial genetic diversity;
- Increased temporal genetic diversity due to the need to replace cultivars with new ones every few years; and
- Increased use of external inputs.

It is crucial to note that monocultural farming methods do not necessarily degrade biodiversity. Indeed, if they take the place of more biologically varied ecosystems, biodiversity loss could result. However, pressure to open up biologically varied ecosystems to cultivation may be lessened if a monocultural system yields more harvests annually and/or greater yields per harvest than the more poly-cultural agro-ecosystem it replaced. According to Kothari and Anuradha131, IPRs by themselves cannot be held accountable for the decline in agro-biodiversity; rather, they will inevitably promote the eradication of a large variety of traditional local varieties in favor of a few widely adapted hybrids and uniform modern varieties. Furthermore, they note that one of the lessons learnt from the Green Revolution is that the seed industry's efforts to create new varieties are unlikely to be equal to the decline of old kinds following their introduction. However, the expansion of monocultural systems won't always lead to the loss of biodiversity. Pressure to open up biodiverse ecosystems to cultivation may be lessened if a monocultural system yields more harvests annually and/or higher yields per harvest than the more poly-cultural agro-ecosystem it replaced (but the opposite result is also possible). However, it's equally critical to note that this tendency in crop breeding started with the start of the Green Revolution and even earlier in some nations. Public crop breeding organisations, not private companies, created the cultivars most frequently linked to the Green Revolution. This implies that, on the surface, this could not even be an IPR issue. Other measures that could promote the adoption of new crop types and the extinction of landraces include the following, according a draft report prepared for the CBD Secretariat for the third COP132 meeting:

- Government farm credits and subsidies, and extension services;
- The policies and programs of international agencies and donor institutions;
- The marketing and research and development policies and programs of transnational corporations; and
- The increasingly concentrated corporate control of pesticide and agro-biotechnology research and distribution.

Plant genetic resources for food and agriculture (PGRFA) should be preserved as "the basis of natural and directed evolution in the plant species most critical to the survival and wellbeing of human beings," according to the Leipzig Declaration on conservation and sustainable utilization of plant genetic resources for food and agriculture, which was adopted by the International Technical Conference on Plant Genetic Resources on June 23, 1996. To face the associated problems of environmental changes, including climate change, and to sustainably expand food supplies and agricultural production, all nations need plant genetic resources. According to Rangnekar, PVP gives "juridical legitimization to the breeding of genetically uniform varieties" and promotes plant breedin□ based on material that is already in scientific use. It is ar□ued that the breeders' exemption, which allows the unrestricted use of plant genetic resources that are already in use, does little to promote the input and discovery of resources that mi□ht be present in the fields of traditional cultivators and other ecosystem types with comparatively high biodiversity levels. PVP proponents can argue that more varieties were likely introduced into the European and Northern markets than they would have been in the absence of a PVP system's incentive. However, a greater number of plant types grown does not always translate into a higher level of agro-biodiversity than would be found in farmers' fields otherwise. This is because plant breeders typically do not claim exclusionary rights over discovered genes or plants into which they are put, and new varieties are often founded on the recombination of genes obtained from a very small gene pool shared by plant breeders. Additionally, according to Rangnekar, PVP rights incentivize breeders to use planned obsolescence techniques "to reduce the durability of plant varieties so that farmers can regularly purchase replacements." He cites empirical data to support his contention that UK wheat breeders do use these tactics. The substitution of better species and varieties for indigenous ones, which lack the varied genetic endowment of the traditional farmers' kinds, is one of the main factors that has been linked to the loss of crop variety. There have been reports of genetic degradation in both industrialised and underdeveloped nations. For instance, according to the FAO, just 20% of the 1930 native Mexican maize varieties are still in use today. Similarly, between 1949 and 1970, the number of wheat varieties in China dropped by a factor of 10.

The current trend of seed corporations to create hybrids and other contemporary varieties that rely on the use of agrochemicals (such as fertilizers, herbicides, and insecticides) to attain high yields has a substantial impact on conservation as well. One typical charge is that these pesticides are used excessively, which kills other plants that grow nearby. However, this trend in crop breeding, which began with the Green Revolution and even earlier in certain countries, is unlikely to be directly caused by IPRs. The development and spread of high-yielding wheat and rice varieties that required extensive agrochemical applications during the 1950s and 1960s are often attributed to the Green Revolution; however, the varieties most commonly linked to the Green Revolution were created by public crop breeding institutions and were not protected by intellectual property ri□hts. In the case of □enetically modified crops, however, the IPR connection seems to be greater. Transgenic plants with inherent resistance to insect pests or herbicides marketed by the same company135 have become more and more common in recent years thanks to life-science businesses, which were formerly chemical companies that acquired seed companies. In the former scenario, it is likely that the herbicide and the seed for which it is intended will both be protected by a patent. For instance, Monsanto was worried that when rival manufacturers of the same herbicide entered the market, it wouldn't experience a sharp decline in sales because it had made substantial profits from one of its patented a□rochemicals, a □lyphosate-based herbicide sold under the brand name Roundup, after the patent expired. For an answer, Monsanto looked to biotechnology. The business created and patented transgenic corn, cotton, soybeans and canola that carried a gene that made them resistant to its Roundup. The Roundup resistance gene and any plants that have it are protected by Monsanto's patents, which are still valid for a number of years. Sales of the seeds are beneficial for sales of the herbicides, and vice versa, because farmers who purchase these "Roundup Ready" seeds are contractually required to buy Monsanto's patented herbicides. The long-term viability of this approach is uncertain, though. Since Roundup Ultra's patent expired in 2000, farmers may choose to use less expensive brands that are offered by rival companies. Monsanto's New-Leaf potato, which promises complete protection against the Colorado beetle, is an example of a crop that has inherent pest resistance (as opposed to herbicide or pesticide). Another is the proprietary Bt maize from Novartis, which is made to withstand the pest of the European maize borer. Large life-science companies like Novartis and Monsanto hold the view that genetic engineering can eliminate or drastically cut down on the use of pesticides. According to Monsanto, their goal is not to force farmers to use more herbicides when they create packages of herbicides and plants that are resistant to them. Making ensuring farmers apply their herbicides is their first concern. The overall use of herbicides may decline if these prove to be more successful than substitute □oods. The firm claims that "roundup herbicide can help reduce tillage to conserve soil moisture and reduce erosion of valuable topsoil, as well as reduce the number of weed treatments." Some scientists and environmentalists ar□ue that □enetically modified herbicide resistance has detrimental impacts on the environment. The following are some of the frequently asserted benefits of usin□herbicide-resistant trans□enic plants:

- Encourage excessive use of herbicides which may kill other plant varieties and species;
- Accelerate the development of resistance among pests; and
- Create the possibility of herbicide resistant genes crossing over to other plants including the weeds being targeted.

This might produce "superweeds," which would eventually make the pesticide useless, and have unpredictable ecological effects. Although this is improbable in the case of the most extensively domesticated crop species, it is also feasible that transgenic plants themselves could turn into "weeds" if the new trait gives them a competitive edge over nearby wild species. Additionally, there are worries that the growing usage of hybrids and other contemporary cultivars created expressly for use with other proprietary agricultural inputs like pesticides and fertilizers could have detrimental social effects, particularly in developing nations. These connections between pesticides, herbicides, and crops might be seen as a move towards capital-intensive agriculture, which raises farming expenses and could be harmful to small farmers. Therefore, detractors contend that farmers shouldn't be the targets of aggressive sales promotion activities and should have the freedom to decide whether or not to accept these packages. Are IPRs involved just because plants (whether transgenic or not), herbicides, and pesticides can be patented, even if we acknowledge that these worries are legitimate? Businesses in these tech sectors frequently assert that they would not be motivated to innovate or create new things if they were not protected by intellectual property rights. This implies that without IPRs, these products would not be available. However, this does not imply that the national patent office should handle such products' marketing approval. To determine whether or not plant herbicide-pesticide packages are in the public interest, the majority of nations have an agency with jurisdiction over such topics, and this body is likely far more qualified than the patent office. In summary, there is a lack of solid empirical data about the relationship between IPR and genetic erosion. One thing that can be assumed with some degree of certainty is that there is no one cause for the loss of agro-biodiversity. Several proximate and underlying factors were identified in one investigation of the connections between agriculture and biodiversity degradation, but IPRs were left out (Table 7.1).

Table 7.1. *Causes of Biodiversity Losses Linked to Agriculture*

Problems	**Proximate Causes**	**Underlying Causes (for all Problems)**
Erosion of genetic resources (livestock and plants) Leads to disease/insect pests Loss of insect diversity	Spread of High Yielding Varieties (HYVs) and monocultures Biases in breeding methods Weak conservation methods	Demographic changes Industrial/Green Revolution Model that stresses uniformity Disparities in resource distribution and in control of land Pressures and influences of seed/agrochemical companies Policies that support HYVs, uniformity and chemicals (subsidies, credit, market standards) Producers/companies focus on short-term returns to neglect of longer-term social factors Disrespect for local knowledge and structural inequities
Erosion of insect diversity	Heavy use of pesticides Use of monocultures Loss of organic material	Policies that support HYVs, uniformity and chemicals (subsidies, credit, market standards) Demographic changes
Erosion of soil diversity Leads to fertility loss Productivity decline	Heavy use of agrochemicals Poor tillage practices Use of monocultures	Policies that support HYVs, uniformity and chemicals (subsidies, credit, market standards) Demographic changes

Erosion of habitat diversity (social and private costs)	Extensification in marginal land Drift/spillover from chemicals	Demographic changes
Erosion of indigenous methods for using agro-diversity	Replacement by uniform species	Disrespect for local knowledge and structural inequities

Source: Thrupp (2002).

7.1.8. Technology Transfer and Food Security

Several developing nations have pointed out the conflict between the TRIPS Agreement's goals of knowledge transfer and development and the manner it allowed rights holders to impose irrational conditions on innovations. Article 7 of the TRIPS Agreement outlines its goals in the following terms: "The defense and upholding of intellectual property rights ought to support technological innovation, transfer, and dissemination in a way that benefits both technology producers and users, promotes social and economic well-being, and maintains a balance between rights and responsibilities." The connection between development and the TRIPS Agreement has been discussed both narrowly in relation to the Agreement's implementation and more broadly in reference to the human rights to nutrition and health. In order for poor nations to handle the effects of IPRs on biotechnological research, they must strengthen their capacity. Jackson suggests creating a Genetic Resource and International Trade Institute in Geneva "to provide developing countries with technical assistance, training, and research on genetic resources management and the rapidly changing policy environment." Article 16 of the Convention on Biological Diversity (CBD) permits the transfer and availability of biotechnology. According to Article 16.2, access to and transfer of technology to developing nations must be made on fair and advantageous terms, including on concessional and preferential terms when both parties agree. For the benefit of both governmental institutions and the private sector of developing nations, each Contracting Party commits in Article 16.4 to "take legislative, administrative, or policy measures, as appropriate, with the aim that the private sector facilitates access to, joint development, and transfer of technology." According to Article 17.1 of the CBD, the Contracting Parties must facilitate "the exchange of information, from all publicly available sources, relevant to the conservation and sustainable use of biological diversity, taking account the special needs of developing countries." Article 17.2 further states that this information exchange must include "information on training and surveying programs, specialized knowledge, indigenous and traditional knowledge as such, as well as results of technical, scientific, and socio-economic research." "Promote international technical and scientific cooperation in the field of conservation and sustainable use of biological diversity, where necessary, through the appropriate international and national institutions," according to Article 18 of the CBD, mandates the Contracting Parties. Article 18.4 envisions the promotion of cooperation in the training of personnel and exchange of experts, while Article 18.5 calls for the "establishment of joint research programs and joint ventures for the development of technologies" pertinent to the CBD's goals. Article 18.2 also mandates that "special attention should be given to the development and strengthening of national capabilities, by means of human resources development and institution building." There is little proof that affluent nations have systematically executed these provisions. These technology transfer responsibilities are, nevertheless, linked to the requirements for bioprospecting permits in several developing nations. "In-Bio" was established as a public organization in Costa Rica to give the country access to biotechnology resources in the form of technological licenses and to make its genetic resources and biotechnology assets available for commercial use. This offers a preliminary example of how to use biotechnology intellectual property assets in a way that is equitable, thorough, and consultative. To share research findings, some public research organizations and universities in both developed and developing nations have set up technology transfer departments. Evaluating the institution's research efforts and identifying business partners who will license the technology and help commercialize research findings are frequently the main goals of the technology transfer office.

7.1.9. Geographical Indications for Agricultural Products

The discussion in the WTO's TRIPS Council regarding the potential expansion of the suggested system for the multilateral registration of wines to other goods was mentioned above. This will be especially important for agricultural product protection, if the European experience provides any indication. As previously mentioned, the European Commission has issued a set of rules pertaining to wine and spirit classifications.141 The two most significant of them are Council Regulation 2082/92 on certificates of specific character for agricultural products and foodstuffs and Council Regulation 2081/92 on the protection of geographical indications and designations for agricultural products and foodstuffs. The protection of geographical indications and designations of origin for agricultural goods meant for human consumption is governed by these regulations. They distinguish between Protected Geographic Indications (PGI) and Protected Designations of Origin (PDO). "The name of a region, a specific place, or, in exceptional cases, a country, used to describe an agricultural product or a foodstuff originating in that region, specific place, or country, and the quality or characteristics of which are essentially or exclusively due to a particular geographical environment with its inherent natural and human factors, and the production, processing, and preparation of which take place in the defined geographical area" is the definition of a designation of origin. A food item or agricultural product must meet certain requirements in order to be eligible to utilise a PDO. These requirements include:

- The name of the food item or agricultural product, along with the place of origin;
- A description of the culinary or agricultural product that includes the main physical, chemical, microbiological, and/or organoleptic properties as well as the raw materials, if applicable; evidence that the food or agricultural product comes from the geographical area; the definition of the ￼eo￼raphical area;
- A description of how the food or agricultural product is obtained, together with, if applicable, the genuine and consistent local methods;
- The details bearing out the link with the geographical environment or the geographical origin;
- Details of the inspection structures; and
- The precise labelling information pertaining to the PDO designation or its conventional country equivalents.

In Case 12–74, Commission of the European Communities v. the Federal Republic of Germany144, the European Court of Justice (ECJ) addressed the issue of what constitutes a registrable PDO. The case involved an application by the Federal Republic of Germany to reserve the appellations "Sekt" and "Weinbrand" for the domestic product and "Praedikatssekt" for wines made in Germany using a minimum amount of German grapes. As German consumers and other interested parties did not understand the appellation 'Praedikatssekt' to describe a wine produced from a fixed minimum proportion of German grapes, but simply 'Sekt' of a particular quality, the Commission contended that the appellations 'Sekt' and 'Weinbrand' were generic appellations that the German legislature attempted to transform into indirect indications of origin through a legislative measure. Additionally, it said that by requiring a minimum percentage of German grapes to be used for these appellations, the law violated Article 30 of the EC Treaty by enacting measures that had the same effect as quantitative limits. The Court pointed out that the German laws pertaining to wine products stipulate that the terms "Sekt" and "Weinbrand" must be used to refer to goods that are either native to the Federal Republic of Germany or imported from other nations where German is the official lan￼ua￼e. Given that the products in question might be made from grapes of unknown origin, it decided that "an area of ori□in which is defined on the basis either of the extent of national territory or a linguistic criterion cannot constitute a geographical area capable of justifying an indication of origin." In the case of vine products, the Court noted that the natural characteristics of the region of origin, such as the grape from which these products were derived, were crucial in determining their quality and characteristics. It also noted that, while the production method employed for these products may have some influence on their characteristics, it was not the only factor that determined their origin, independent of the quality of the grape used. The Court came to the conclusion that it was not possible to demonstrate that the products in question have unique qualities and traits that made them characteristically German. In order to demonstrate that German consumers thought of the appellations "Sekt" and "Weinbrand" to relate to domestic German wines, the defendant had provided opinion polls. This was dismissed by the Court because the protection provided by the indication of ori□in could only be justified if the product in question truly had qualities that might set it apart from its place of origin. Without such a requirement, this protection could not be supported by consumer opinions, such as those derived from surveys conducted using statistical standards.

The Court came to the conclusion that the German laws pertaining to vine products included provisions that, in violation of Article 12(2)(b) of Regulation No. 816/70, had the same impact as quantitative import limits. "The name of a re□ion, a specific place, or, in exceptional cases, a country, used to describe an agricultural product or a foodstuff" is what is known as a PGI.

- Originating in that region, location, or nation, and having a certain quality, reputation, or other attributes that are distinctive to that geographic origin, as well as production, processing, and/or preparation occurring inside the designated geographic area.

The issue of PGIs was examined in Case C-269/99145, which involved a 26 January 1994 application for gherkins originating in the region known as "the glacial valley of the Spree between the northern edge of the town of Cottbus and the Neuendorfer See, north of the town of Lübben" submitted by the German Government under Article 17 of Regulation No 2081/92. The application to register the designation "Spreewälder Gurken" was met with opposition from several interested third parties. They asserted that the processed product should not contain any raw materials from other production areas and that the unique geographical and climatic conditions mentioned in the application only applied to the Spreewald, strictly speaking, in the inland delta region and not to the entire Wirtschaftsraum Spreewald. The ECJ held that national courts should decide whether a designation application was legal. Similarly, the Commission could legitimately register the designation "Spreewälder Gurken" under the streamlined method because the national court had made its own evaluation, which was untarnished by a glaring error. Because a food product may be considered to originate from the geographical area in question if it is processed or made there, even if the raw materials are produced elsewhere, the designation "Spreewälder Gurken" was registered as a PGI. The registration of the "Spreewälder Gurken" desi□nation for the re□ion specified in the revised specification was based on the same idea. Names that have become generic may not be registered, according to EU regulations. All of the following considerations must be considered when determining whether a name has become generic:

- The existing situation in the Member State in which the name originates and in areas of consumption;
- The existing situation in other Member States; and
- The relevant national or Community laws.

Many of the products produced by agricultural communities in developing nations will have become generic due to their general lack of awareness with the notion of regional indications. In Europe, this may have been the case, as evidenced by the cheese's name, Feta. For "salted white cheese traditionally produced in Greece, from sheep's milk or a mixture of sheep's milk and goats' milk coming exclusively from the regions of Macedonia, Thrace, Epirus, Thessaly, Central Greece, Peloponnese, and Lesbos," Greece had applied to register "Feta" as a PDO. Despite the fact that the majority of Member States had requested that the Commission add the name "Feta" to the list of generic names it was creating. According to the Commission, "Feta" was not excluded from registration for this reason. It had depended on a Greek market assessment that found Feta was accepted as a geographical designation in that nation. The ECJ held that the Commission could not downplay the si□nificance of the situation in Member States other than the State of Origin and that consideration should be given to the existence of products sold legally under that name in Member States other than the State of origin where registration is being sought. The ECJ decided that the disputed regulation had to be revoked to the degree that it registered the name "Feta" as a PDO because the Commission failed to adequately evaluate all of the elements that Article 3(1) of the basic regulation required it to consider. This kind of situation demonstrates how crucial it is for governments to provide unique geographic indications to support the sale of their agricultural goods and prevent genericity. In situations where generic phrases have become commonplace, they can be linked to specific □eo□raphic indicators. For instance, Macedonian Rakia or Bulgarian Rakia would distinguish between manufacturers of the spirit drink that is made all throughout the Balkan Peninsula. Food processing and packaging as part of a PDO under Regulation

No 2081/92 have been the subject of several recent ECJ cases. In Case C-47/90, the PDO "Rioja" was involved. Wine made in the La Rioja region was given a "denominación de origen" under Law No. 25/70, Article 84 et seq. A Governing Council for the Rioja designation of origin was established at that time. Rules pertainin□ to the classification and the Rioja were authorized by the Spanish Minister of Agriculture, Fisheries, and Food on April 3, 1991146. According to Article 32 of the Rioja Rules, wines protected by the denominación de origen calificada Rioja may only be put into circulation and shipped from re□istered cellars in special bottles that do not diminish their quality or prestige and have been approved by the Governing Council. Additionally, the wine may only be bottled in registered cellars authorized by the Governing Council; otherwise, it may not bear that designation. Except for one-liter bottles, the bottles must be made of glass and have a capacity permitted by the European Economic Community.

The Commission and the governments of Spain, Italy, and Portugal presented fresh evidence to show that the arguments supporting the disputed requirement could support it. They contended that community legislation generally tends to improve product quality within the parameters of the common agricultural policy, especially in the wine industry. This is done, among other things, by using designations of origin that are given special protection. The Court acknowledged that the unique traits and characteristics of Rioja wine are the consequence of a confluence of human and natural forces related to its geographical origin, and that measures must be made to ensure their preservation. It acknowledged that certain traits and attributes were intended to be upheld by the re□ulations controllin□ the Rioja "denominación de ori□en calificada." It was acknowledged that the regulation of bottling was an attempt to better protect the product's quality and, in turn, the designation's reputation. Without this criterion, the presti□e of the Rioja "denominación de ori□en calificada" could actually be weakened, according to the Spanish government, which is backed by the governments of Italy and Portugal as well as the Commission. It believes that shipping and bottling the wine outside of its production area would compromise its quality. The court acknowledged that wine bottling is a si□nificant process that could si□nificantly lower product quality if not done in compliance with strin□ent □uidelines. Bottlin□ is more than just fillin□ empty bottles; it typically involves a number of intricate oenolo□ical processes (filterin□, clarifying, cooling, and so forth) that, if not completed in compliance with industry standards, could degrade the wine's quality and change its properties. It was acknowledged that, if not done under ideal circumstances, bulk wine transportation mi□ht si□nificantly de□rade its quality. In the case of bottlin□, the best conditions are more likely to be guaranteed if the bottling is carried out by undertakings established in the region of those entitled to use the designation and operating under their direct control, as they have specialized experience and, more importantly, thorou□h knowled□e of the specific characteristics of the wine in question that must not be impaired or lost at the time of bottling. This is despite the Court's acceptance that, under the best circumstances, a wine's characteristics and quality may indeed be maintained when it has been transported in bulk and bottled outside the region of production. It can be inferred that the risk to which the quality of the product ultimately offered to consumers is exposed is greater when it is transported and bottled outside the region of production than when those operations have taken place within it. The Court noted that the controls for Rioja wines that are transported and bottled in the region of production are extensive and systematic, and they are the responsibility of the producers themselves. The requirement in question was therefore deemed to be justified as a measure to protect the "denominación de ori□en calificada which may be used by all the producers concerned and is of decisive importance to them." The requirement was designed to strengthen control over the unique characteristics and quality of Rioja wine in order to preserve its considerable reputation.

Case C-108/01147 dealt with Italian Law No. 26 on protection of the designation of origin 'Prosciutto di Parma' of February 13, 1990, which reserved the designation 'Prosciutto di Parma' ('Parma ham') exclusively for ham prepared from fresh legs of pigs raised and slaughtered in mainland Italy, produced in compliance with the law's provisions, and marked with a distin□uishin□ mark that allows it to be identified at any time. Accordin□ to the law, Parma Ham must only be sliced and packaged at facilities in the usual production area that have been authorized by the Consorzio, which is in charge of overseeing Parma Ham production. In the case, Asda Stores Ltd., a UK supermarket chain, sold ham with the name "Parma ham" that it had bought pre-sliced from Hygrade Foods Ltd., which in turn bought the ham boned but not sliced from an Italian manufacturer that was a member of the Consorzio. Hygrade chopped the ham and hermetically wrapped it in packets that read "ASDA." Italian-inspired PARMA HAM Real Parma Ham from Italy. On the grounds that Asda and Hygrade were in violation of the regulations governing Parma Ham, the Consorzio filed a writ in the UK to obtain a number of injunctions against them, essentially ordering them to stop their operations. The House of Lords asked the ECJ to rule on whether the Italian law, Commission Re□ulation (EC) No 1107/96, and the PDO Prosciutto di Parma specification established a legitimate Community right that could be directly enforced in a Member State court to prohibit the retail sale of sliced and packaged ham that was derived from hams that were legally exported from Parma in accordance with the PDO's conditions but had not been sliced, packaged and labelled in accordance Applying the "Rioja" case, the Court decided that Regulation No.

2081/92 should be read to allow the use of a PDO subject to the requirement that certain operations, like product slicing and packaging, occur in the production re□ion, as specified in the specification. The Court acknowled□ed that this measure was justified for ensurin□the authenticity and quality of the product, even though it had the same effect as a quantitative export restriction. It pointed out that Community legislation generally tended to improve product quality within the framework of the common agricultural policy, promoting the reputation of those products through the use of designations of origin that enjoy special protection, among other things. As industrial and commercial property rights, indications of origin shielded their rightful owners against unauthorized usage by those parties looking to capitalize on their established reputation. As per the ruling, the PDO 'Prosciutto di Parma' standard, which mandates that the slicing and packing be done in the producing region, was designed to provide the people who are allowed to use the PDO control over one of the ways the product is marketed. This requirement was defended on the grounds that it protects the product's quality and validity, and hence the PDO's reputation, which belongs to the people who are allowed to use it. The Court pointed out that the slicing and packin□ of Parma ham were si□nificant activities that may dama□e the product's quality and, consequently, the PDO's reputation if they were done in a way that left the final product without the anticipated or□anoleptic properties. Since those procedures inevitably entail removing the whole hams' indication of origin, they may also jeopardize the product's authenticity guarantee. In order to protect the reputation of Parma ham by enhancing control over its unique qualities and characteristics, the Court decided that the requirement of slicing and packa□in□in the production re□ion may be justified as a measure protectin□ the PDO, which is of crucial importance to all the operators involved. Since there were no less restricted options that could achieve the goal, the resulting restriction was consequently seen as essential to achieving it. Ultimately, it came to the conclusion that the PDO "Prosciutto di Parma" would not be afforded the same level of protection from a requirement placed on operators based outside the production region to notify customers—through suitable labeling—that the slicing and packaging was done outside of that region. Therefore, the reputation of all gammon marketed under the PDO, including that sliced and packaged in the region of production under the control of the group of producers entitled to use the PDO, could be harmed by any decline in the quality or authenticity of gammon sliced and packaged outside the region of production that results from the materialization of the risks associated with slicing and packaging. Infringement cases pertaining to geographical indications typically involve one of the following:

- Wrongful use of a PGI or a PDO, in which case an action will be brought by the entity responsible for preserving the integrity of the geographical indication; or

- In relation to a misleading use of a geographical indication.

India, a developing nation that enacted the Geographical indicators of Goods (Registration and Protection) Act in 1999, has indicated that it wants to safeguard its agricultural products through the use of geographical indicators. "Any agricultural, natural or manufactured goods or any goods of handicraft or of industry and [it] includes foodstuff" is how the Act defines "□oods" (Article 2 (1) f). Accordin□to the Act, a name that is used in connection with specific goods that originate from a particular area but does not designate a geographical area can also be regarded as a geographical indication, provided that the name is not generic or descriptive of the product (as to the kind, nature, type, etc. of the product). The purpose of such a definition was to safe□uard "Basmati" designations. The Act reproduces TRIPs Art. 22 (4) verbatim and prohibits the registration of indications that, while accurately describing the territory, region, or locality in which the goods originate, deceive the public into believing that the goods originate in another territory. More generally, it is illegal to use false geographical markings that imply that the goods are made in a different region than their actual place of origin. are also considered to be violations of the use of geographical indications in a way that amounts to unfair competition, which is any conduct that denigrates or tarnishes a rival. Any organisation of individuals or producers listed in the registry as the proprietor of the geographical indicator is awarded registration for a period of 10 years, renewed following a period of opposition. Due to its openness to international groups, this protection structure is comparable to that established for trademarks. It does, however, have certain characteristics unique to regimes governing geographical indications, such as the authority the Registrar of Industry Standards has over the production, manufacturing, and exploitation of goods with particular qualities, reputations, or other attributes that are primarily due to their "geographical origin with the detailed description of the human creativity involved." Another characteristic of a regime of protection of geographical indications is the method to guarantee that the producers maintain the criteria. In addition, collective and certification trademarks are provided by the Indian Trade Marks Act 47 of 1999. "The ability to differentiate the goods or services in connection with which it is used in the course of trade which are certified by the mark proprietor in respect of ori□in, material, mode of manufacture of goods or performance of services quality" is a requirement for a certification trademark. The term "collective mark" refers to a trademark that sets apart the products or services of members of a group of people who own the mark from those of other parties. The rules governing the use of a collective mark must be submitted with an application for a collective trademark. These rules must outline the requirements for association membership, who is permitted to use the mark, and how the mark is to be used. "If a collective mark is likely to mislead or confuse the public, especially if it is likely to be mistaken for something other than a collective mark, it shall not be registered." A draft of the rules controlling its usage must be submitted with an application for a certification trademark. When the owner is no lon□er qualified to certify the □oods or services in question, or when "it is no lon□er in the public advantage that the mark should remain registered," the registration may be challenged, withdrawn, or altered.

7.1.10. Terminator Technology

As the name implies, the term "terminator technology" was not created by supporters but rather by a Canadian activist named Pat Mooney, who wanted to draw attention to it negatively. He was very effective in this. Mooney first became aware of Terminator in 1998 after he saw an announcement that Delta and Pine Land, a significant American cotton seed company, and the US Department of Agriculture had jointly received a patent148 that described molecular biological methods for regulating gene expression in plants, plant parts, or seeds so that traits could be turned on and off between generations. Depending on the qualities in question, whose expression or lack thereof may contribute to the harvest's success, farmers can perhaps profit from these methods. However, a technique for creating seed that cannot germinate—more precisely, a technology that would make harvested seed sterile—is one of the promises. On the surface, it appears remarkable that so much time and money would be spent creating a method of producing sterile seed. However, this is purely commercial in nature, even though a public sector organisation is involved. The goal is to undermine seed business monopolies by preventing farmers from replanting stored seed. By doing this, it offers a way to guarantee the continuation of the monopoly after the expiration of any patent or plant variety certificate, assuming that such actions call for the consent of the relevant right holder, in addition to preventing the infringement of intellectual property protection. Furthermore, terminator technology has serious ramifications for breeding, which depends on unfettered access to plant kinds for the purpose of generating initial variety. The growing use of licensing agreements that require customer-farmers to refrain from replanting their patent-protected seeds also appears to be a reflection of the private sector's increased determination to stop the replanting of proprietary seeds, which is evident in the development of this technology (and, as we shall see below, in the case of hybrids, with the help of a public agency). Naturally, if this technology were to become extensively employed, such agreements would no longer be required. Terminator is one of many genetic use restriction technologies (GURTs), which are not new. By crossing inbred pure lines, a US public sector plant scientist named George Shull identified the phenomena of (what he called) "heterosis" in maize plants around the beginning of the 20th century. Increased yields are the result of this process, which is often called "hybrid vigor." However, because they are hybrids, the yield boosts are only permanent for one generation because the progeny cannot breed true. Therefore, even though farmers stand to gain from seeds that offer this hybrid vigor, they must purchase seeds at the start of each planting season in order to reap the benefits of equally fruitful harvests in the future. The seed firms benefited from this requirement and still do. Farmers will keep using hybrids instead of their open pollinating counterparts as long as the extra money made from the hybrid vigor outweighs the expense of buying seeds. Rangnekar lists four strategies used in a discussion of private appropriation in the plant breeding industry. These are

- IPRs and seed market regulation;
- Organizational solutions;
- Discontinuous heritability; and
- Planned obsolescence. He classes hybrid and terminator technologies as methods of discontinuous heritability.

It is widely believed that the hybrid approach to producing superior seeds will benefit the □rowth of the seed industry and, in the view of many, but not all, farmers as well. Actually, a number of the bi□□est seed firms in the world in the 20th century rose to prominence by successfully developing hybrid maize varieties. These consist of Funk, DeKalb, Pfister, and Pioneer Hi-Bred. However, detractors contend that the enormous sums of money spent in the 1920s and 1930s on the creation of hybrid varieties could have been used for breeding using more traditional methods that would have produced comparable yield increases without limiting farmers' ability to replant their harvested seeds. According to Berlan and Lewontin, hybridisation is a type of "deterioration technique" that not only allows seed firms to remove on-farm saving and exchange, but also completely destroys any chance for selective breedin□ to improve crops. Althou□h hybrid cultivars may benefit farmers in the short term, their long-term interests may not always be best served by their broad adoption. Initially, hybrids were not si□nificantly more productive than their conventionally bred relatives. But starting in the middle of the century, more private investment was si□nificantly raisin□hybrid maize yields. Unfortunately for breeders, hybridisation is less commercially viable and does not work as well for some of the most extensively grown crops, such as rice and wheat. Farmers may replant harvested grain as seed or even multiply seed to sell in competition with the breeder (thinkin□this would be more profitable for them than selling harvested produce) because there was no rule prohibiting it. Terminator technology seems to be the answer to the issue. However, terminator, which bases its use restriction on seed sterility, offers no agronomic or productivity benefits to the farmer purchasin□ the seed, in contrast to this previous biotechnology protection system and other GURTs in development that aim only to regulate the expression of particular traits. Even worse, it is a net loss since, unlike hybrid maize and the so-called trait-specific GURTs, it takes away a freedom without providin□ any compensatin□ benefits. Some si□nificant questions are raised by this moment. The most fundamental question is whether the Terminator dispute has any bearing on agriculture's future or if it is all nonsense. Second, is terminator the type of invention we should be supporting in this way if the goal of patenting is to encourage creative endeavour for the public good? The third question is whether there are good enough legal reasons to deny it patent protection.

Lastly, should nations broaden the scope of the order public and morality exclusions that are accessible to World Trade Organisation member states through Article 27.2 of TRIPS, which is interpreted rather narrowly in Europe, in order to eliminate the protection of such technologies if these grounds are absent? The fact that the US Department of Agriculture has been working with the private sector to create GURTs as part of the government's larger, longer-term initiative to safeguard its companies' intellectual property in foreign markets, including developin□nations, may be the primary indication of terminator's si□nificance. In fact, a USDA spokesperson stated that the goal is to "increase the value of proprietary seed owned by US seed companies and to open up markets in Second and Third World countries" by allowing the technology to be "widely licensed and made expeditiously available to many seed companies." Terminator offers a few possible advantages. For starters, it might ease some of the worries of those who oppose □enetically modified crops. These worries include the possibility that genes from these plants could spread to other species, a phenomenon known as horizontal gene transfer that, ironically, many proponents of GM agriculture brush off as unimportant anyhow. Furthermore, safe protection may promote additional investment in plant breeding and agricultural biotechnology, notably in areas that help small farmers. Whereas the latter is wholly theoretical, the former is believable. Terminator's limitations on seed replanting, trade, diffusion, and on-farm breeding may be its worst drawback, assuming for a minute that the technology will promote research focused on small farmers. It is crucial to know that subsistence farmers in underdeveloped nations typically obtain their seeds from their own or their neighbors' farms in order to explain why. It is crucial to keep the freedom to do this for two reasons. First, at the beginning of each planting season, subsistence farmers frequently lack the money or credit necessary to purchase seed. Purchasin□ seed is a si□nificant financial commitment for them. Farmin□ may become even riskier for the impoverished if it turns out that the advantages of terminator-protected seed are not enough to offset its increased cost. Although farmers can probably go back to their traditional crops, desperation could result from a single bad harvest combined with mounting debt. Second, small-scale farmers in many developing nations do much more than just plant seeds that have been grown elsewhere. It is true that local varieties are the product of generations of improvement through trial and selection on the farm; in the present day, these methods may entail current kinds that may need to be modified to fit local conditions. Such abuses can be stopped by making these farmers merely clients of businesses that offer terminator-protected seed. In addition to potentially harming local food security, this could undermine global plant breeding efforts by lowering the variety of germplasm available if it spreads. For these reasons, Collins' claim that "the centuries-old practice of farmer saved seed is really a gross disadvantage to Third World farmers who inadvertently become locked into obsolete varieties because of their taking the "easy road" and not planting newer, more productive varieties" is unlikely to resonate with or provide much solace to subsistence farmers. Farming communities may become impoverished if GURT-protected seed yields are subpar and if it requires more inputs, such as agrochemicals, than traditional varieties (which is frequently the case with current varieties).

Advocates will undoubtedly contend that farmers still have options. However, in practice, they might not. The ability to patent such a technology is viewed by many detractors as a critique of the patent system. It is reasonable to wonder if society should support this kind of study by offering a patent monopoly. Furthermore, it is reasonable to worry that patenting and GURT-protecting seeds is overprotective, much as copyright law's support for encryption by prohibiting circumvention devices is unduly beneficial to owners. Furthermore, certain nations may view such technologies as immoral or against the public order, and this is a decision that should be honored. Since they have the authority to establish their own standards for what constitutes immorality or ordre public, they are legally secure. Terminator has already been outlawed in India, and it is conceivable that other developing nations would follow suit. Terminators are obviously not the kind of technology that these nations aim to promote through their patent systems. Nonetheless, it might be argued that expanding the use of these exclusions is not specifically necessary in response to terminator-type patents. Indeed, terminator mi□ht stimulate further study in this field if it were not patentable on these or other grounds. After all, it would seem that GURTs are particularly helpful in areas with inadequate IPR protection. Furthermore, a patent does not grant the right to economic exploitation. The use of such technologies should not be unrestricted; rather, it should be approved through a procedure based on good science, evaluations of the effects on the environment and society, and perhaps the precautionary principle. Authorities that grant patents may not be equipped to handle all of this. Despite the fact that they have frequently been overlooked in the clamor, the terminator patent highlights two primary difficulties. First, terminator technolo□y is a prime example of how agricultural research is becoming increasingly costly, technologically sophisticated, and focused on commercial interests. As a result, the industry is becoming one in which fewer and fewer businesses can enter, while those that are already present and able to compete grow to dominate it. Indeed, Terminator has the potential to severely undermine public sector research while speeding up the already apparent corporate concentration process. It might accomplish this by strengthening the restrictions on plant genetic resources, forcing others to either forgo them entirely or pay license fees that could be prohibitively expensive for public sector organisations and competitors. If so, nations must enact competition laws that preserve the public sector's autonomy to conduct agricultural research while guaranteeing farmers and consumers a choice. Second, we let the private sector dominate agricultural research at our own risk and the risk of farmers in developing nations in particular, who will inevitably be overlooked in the same manner that pharmaceutical corporations disregard the illnesses of the underprivile□ed for sound financial reasons. Even if GURT-protected seeds are created for impoverished farmers, Terminator might make matters worse for them because it will limit their ability to carry out on-farm breeding experiments, which are frequently required to modify varieties to better suit their unique requirements. If public sector experts in other nations share the USDA's reasons for endorsing Terminator, it is also a very negative thing. Globally, public sector agricultural research is dwindling. However, the need for studies aimed at impoverished farmers has never been greater. As the terminator study clearly shows, not all public sector research accomplishes this, but if industry is left to handle all the research, we can be certain that even less will be accomplished. In fact, for impoverished farmers, the end of public sector research might pose a greater challenge than terminator technology.

7.2. FOOD SECURITY AND IP RELATIONSHIP

In order to eradicate food insecurity, intellectual property must be seen from the appropriate policy standpoint. Food insecurity and rural poverty in emerging nations are attributed to development plans that failed to recognize the significance of strengthening the agricultural sector, especially the production of staple foods, according to development experience during the past 50 years. Therefore, a set of policies addressing the supply, distribution, and consumption facets of the food chain are necessary to improve food security in developing nations. According to the FAO, a variety of constraints limit the policy options open to poor countries, including:

- Limited resources for public spending programs;
- The dilemma between remunerative prices for producers and prices that a large number of poor households can afford, thus making the option of border protection less attractive, despite high bound tariffs; and
- Major constraints on foreign exchange availability leading to pressure to boost production of export crops.

The promotion of advantageous agricultural advances is one area where intellectual property might have the biggest impact. The marketing of hybrid seeds, which provide higher yields and offer the seed marketer the commercial advantage that the farmer cannot utilise the offspring's seeds since they do not breed true to type, has historically produced the highest incentives. As was previously mentioned, there is no evidence to support incentives for crop plant breeding research, and it is even less clear if PVP and patenting will be effective in promoting a domestic seed business in underdeveloped nations. Although there is a chance that several countries with comparable agricultural conditions could combine their markets in a way that promoted private investment, Barton says that a developing nation "is probably best-off adopting minimum compliance with TRIPS, which requires at least some form of sui generis protection for plants." Additionally, commercializing varieties created by the public sector may be aided by the application of UPOV-style rules. The technological sophistication of agricultural research in a developing nation will determine whether it will implement a patent-based system or a sui generis PVP system to comply with Article 27.3(b) of the TRIPS Agreement (unless the Agreement is modified in the manner recommended in the African Group's communication to the TRIPS Council). PVP would be the most likely path in agricultural research including traditional plant breeding. However, patenting becomes a possibility in countries that have established the capacity to conduct microbiological research. The type of protection that is chosen in both nations will rely on the type of registration facilities that are offered. Facilities for test breeding are necessary for a PVP system. Appropriate examination facilities will be needed for patenting. There is the potential for regional cooperation in both situations. If there are regional patent offices or the capacity to search internationally for PCT members, this is most likely more advanced in terms of patenting. The Community Plant Variety Rights Office of the EU is an illustration of how PVP is regionalized.

7.3. ACTION RECOMMENDATIONS

7.3.1. Policy Capacity Building

The preamble of the TRIPS Agreement acknowledges the "special needs of the least-developed country Members in respect of maximum flexibility in the domestic implementation of laws and regulations in order to enable them to create a sound and viable technological base" as well as "the underlying policy objectives of national systems for the protection of intellectual property, including developmental and technological objectives." The protection and enforcement of intellectual property rights "should contribute to the promotion of technological innovation and to the transfer and dissemination of technology, to the mutual advantage of producers and users of technological knowledge and in a manner conducive to social and economic welfare, and to a balance of rights and obligations," according to Article 7 of the TRIPS Agreement. Developing and least-developed WTO members must build their IP policy capabilities as part of the TRIPS review procedures in general. IP is typically a key element in the complicated and multifaceted debate around food technology and the issue of access to new technologies. The European Commission should consider helping poorer nations build their IP policy capacity. "Members may, in formulating or amending their law and regulations, adopt measures necessary to protect public health and nutrition, and to promote the public interest in sectors of vital importance to their socioeconomic and technological development, provided that such measures are consistent with this Agreement," according to Article 8.1 of the TRIPS Agreement. This provision encourages a multidisciplinary approach, but its exact extent is still up for debate. In fact, in addition to legal issues, economic and financial, scientific and technological, ethical and philosophical, medical, agricultural, and cultural factors must now be taken into account when developing IP and food security policies. As IP policies are currently being formulated by international and inter-governmental organizations as diverse as the WTO, WIPO, FAO, WHO, CBD, UNCTAD, UNEP, UNESCO and WHO, developing and least-developed countries which have to engage with this process have to construct IP policy capacity in the areas of: public health (patenting, confidential information, compulsory licensing, parallel importation); food security (patenting and plant variety protection); agricultural research (access to proprietary enabling technologies, development of IP assets; genomics and bio-informatics, bio-prospecting and access to genetic resources); agricultural trade (patenting, plant variety protection, geographical indications); general trade (trademark protection, piracy and counterfeiting, border control of IP rights); technology transfer (approval of technology transactions, technology packaging, control of restrictive licenses, remuneration); the impact of digital technologies (copyright and computer programs, software patenting, communication technologies, domain names, ecommerce, encryption and technological controls, reprographic technologies, electronic rights management); enforcement of IP rights (civil litigation, judicial adjudication, criminal enforcement, alternative dispute resolution, jurisdictional issues); traditional knowledge and folklore; establishment and management of IPRs (patent examination and searching, registration of rights, compulsory licensing). There are numerous government departments and public entities in every nation that must handle public policy concerns brought up by various IP factions. However, there is little experience in developing and least-developed nations when it comes to creating and implementing IP policies. It is crucial to build IP policy capacity in developing and least-developed nations so they can participate more successfully in IP discussions at the national, regional, and global levels. This initiative's goal is to advance approaches to intellectual property that can support important public policy purposes like food security, innovation, creativity, development, and the public dissemination of knowledge and ideas, rather than to promote understanding about IP as an end in itself. This program would help emerging IP policy leaders in developing and least developed nations who are dedicated to reducing poverty and promoting equity and fairness to participate in the:

- Design and implementation of appropriate domestic policies; and
- Process of international IP standard-setting.

There are several ways the EC can help with this capacity-building. In its most basic form, this would include training programs for individuals involved in establishing and implementing policies. These courses could be offered through regional training programs in underdeveloped nations or at suitable EU institutions. To further the public policy interests of developing and least developed countries, particularly the interests of the poor, ethical considerations, and development in those countries, this capacity building initiative could take the form of a global, self-sustaining network of experts, policymakers, and scholars from developing countries who would participate in national and international IP agricultural policy debates. This network might be seen of as a twinning arrangement between institutions in LDCs and developing nations, which would aid in creating a leadership development and training strategy that offers an alternative to current capacity building initiatives. The program would not only concentrate on understanding and implementing current intellectual property rules; it would also assess them and develop food security policies that are pertinent to local and national contexts. The network would:

- Assist the South in establishing national and regional IP-agri policy networks;

- Determine the various regions' and possible participants' policy and substantive priorities.
- Create standards that direct an evaluation of the needs of prospective trainers and recipients for every facet of IP and food security policy capacity building;
- Provide standards for selectin□partners, mentors, beneficiaries, and others who have demonstrated a commitment to the public interest and a desire to participate in IP and food security policy discussions from a public interest standpoint;
- Becomin□a □lobal platform for efficient knowled□e sharin□in the quickly evolvin□field of intellectual property food security policy; and
- Provide effective mentoring for IP food security policy experts in the South.

The EC can offer intermediate forms of assistance, which are also covered below, by helping to create legislation that addresses the regulation of intellectual property aspects of food security. For instance, the Community Plant Variety Rights Regulation, the EU's activities under the Biotechnology Directive, and its knowledge of geographical indications and food security could all be helpful.

7.3.2. FAO International Treaty on Plant Genetic Resources Implementation

In particular, the FAO International Treaty on Plant Genetic Resources for Food and Agriculture's provisions regarding the denial of intellectual property protection for any material transferred within the framework of the multilateral system, as well as the national implementation of Farmers' Rights, should be promoted by the European Commission. The Treaty and its implementation may eventually be linked to several Material Transfer Agreements (MTAs). These consist of:

- The MTA used by the CGIAR Centers to distribute designated germplasm;
- The standard MTA for the Treaty;
- The MTA to be used by Centers for "in trust" materials of crops/species that are not part of the Multilateral System; and
- MTAs used by the CGIAR Centers and others for the transfer of improved germplasm that may incorporate materials from but not itself be part of the germplasm covered by the Multilateral System.

Developin□nations and LDCs could benefit from the EC's technical support in using these MTAs and in helping the Treaty's Governing Body formulate policies.

7.3.3. Convention on Biological Diversity Implementation

Regarding the bioprospecting operations of European businesses and the distribution of profits from the exploitation of those resources to source communities, the European Commission ought to back the development of the principle of informed consent. This assistance ought to materialize:

- In the negotiations within the TRIPS Council concerning the renegotiation of Article 27.3(b) of the TRIPS Agreement;
- Within the COP of the CBD;
- In the discussions within WIPO on the Substantive Patent Law Treaty; and
- As an obligation of national legislation within the EC and elsewhere.

The work being done in several forums around the issue of origin disclosure in PVP and patenting is closely related to this question. Given the length of time between acquisition and usage in breeding programs, genetic material from plants and plant derivatives may originate from a variety of sources, some of which may no longer be traceable due to a lack of documentation. LDCs and developing nations can receive help from the EC in creating a functional county of ori□in system. In order to achieve this, it would be beneficial to look into how much biogenetic resources are used by the relevant industrial sectors. Additionally, it would be helpful to survey businesses and IP practitioners regarding the feasibility of disclosure of origin and how its implementation would impact their research and development.

7.3.4. Protection of Traditional Agricultural Knowledge

The European Commission need to embrace the idea that traditional agricultural knowledge qualifies as a form of intellectual property that can be protected. This assistance could appear as:

- In the discussions within WIPO for a convention concerning the protection of intellectual property;
- As part of the implementation of the FAO Treaty on the Protection of Plant Genetic Resources for Food and Agriculture;
- As part of the discussions within WIPO on the Substantive Patent Law Treaty; and
- As an obligation of national legislation within the EC and elsewhere.

7.3.5. Extend Geographical Indications to Agriculture Products

It has been proposed that emerging nations who possess or may be able to attain a competitive advantage in agricultural products, processed foods, and drinks may be especially interested in geographical indicators. Naturally, these advantages must be balanced against the cost of enforcement measures and the cost of preserving the geographical indication in the nation of origin. Case studies and cost evaluations of the anticipated effects of implementing a European-style registration system can be prepared with helpful assistance from the EC. In the context of sustainable food security, EC support would also be helpful in investigating with developing nations and LDCs how a geographical indications policy may be incorporated into the creation of rural policy. The proposal to add agricultural and other goods to the international register for the geographical indication of wines, which is envisioned in the framework of the talks under Article 24 of the TRIPS Agreement, could also receive support from the EC.

7.3.6. Technology Transfer in Support of Food Security

In developing nations and LDCs, the European Commission ought to assist technology transfer for the in situ and ex situ conservation of land races and traditional food crops. The EC should promote adherence to the duties outlined in Articles 16–18 of the Convention on Biological Diversity, which connect a nation's ability to access its genetic resources to its ability to access knowledge, technologies, and capacity building pertaining to their exploitation. The needs of poor nations in terms of food security could be the main emphasis of this technological transfer. It is clear that the EC already supports the CGIAR's in-situ conservation initiatives. The field of ex situ conservation of genetic resources for agriculture and food still has a lot of work to do. The creation of industry standards for bioprospecting, where technology transfer is incorporated into technology transfer permits as a requirement of access to genetic resources, may be one area of special support.

7.3.7. Implementation of UPOV

The European Commission should support the protection of farmers' rights to conserve and trade seeds as part of countries' obligations to implement plant variety rights protection, whether as sui generis law or as a UPOV-style regulation. Simultaneously, the European Commission ought to advocate for the preservation of the liability exemption for research employing protected types.

7.3.8. Compatibility of PVP and Patenting

By making sure that the patenting of plant genetic components does not extend to the patenting of plants themselves, the European Commission should promote compatibility between the systems for the protection of plant varieties and patents. This would undermine the research exception in PVP laws and jeopardize food security.

7.3.9. Clearing House Mechanism

The creation of the Clearing House Mechanism in the CBD as a project to disseminate information regarding intellectual property applications pertaining to plant genetic resources for food and agriculture globally should have the backing of the EC. Reducing biodiversity loss and fostering just and equitable benefit sharing would result from the creation of an international system for sharing and integrating data on plant genetic resources. Dealing directly with traditional knowledge stakeholders and source countries through the method will reduce transaction costs and promote the utilization of genetic resources by developing nations. Several beneficial tasks would be carried out by the mechanism:

- A repository for national and community registers of indigenous knowledge, which would be maintained under strict obligations of confidentiality;
- A catalogue of knowledge and innovations which are available for sale or licensing, as well as identifying that traditional knowledge which is unavailable;
- A register of legal experts who are available to assist indigenous and traditional communities in such negotiations and in evaluating research proposals;
- Representing the stakeholders in national government and intergovernmental negotiations;
- Monitoring the use, e.g., patenting of traditional knowledge;
- A dispute resolution facility between stakeholders;
- Promulgating industry bioprospecting standards, and contract terms; and
- Engaging in awareness-raising activities.

It has been suggested that a private global bio-collecting society be formed to maintain the clearinghouse mechanism independent of intergovernmental politics, regardless of whether it operates as a private organization or as a component of a government or intergovernmental structure. Research on whether government and public sector involvement in the clearinghouse system should be mandated should be supported by the European Commissio

Bibliography

1. Aghion, P., & Howitt, P. (1992). A model of growth through creative destruction. *Econometrical, 60*(2), 323–351.
2. Alchian, A. A., & Demsetz, H. (1973). The property right paradigm. *The Journal of Economic History, 33*(1), 16–27.
3. Amabile, T. M., & Pratt, M. G. (2016). The dynamic componential model of creativity and innovation in organizations: Making progress, making meaning. *Research in Organizational Behavior, 36*, 157–183.
4. Archibugi, D., & Filippetti, A. (2010). The globalization of intellectual property rights: Four learned lessons and four theses. *Global Policy, 1*(2), 137–149.
5. Arrow, K. J. (1972). Economic welfare and the allocation of resources for invention. In *Readings in Industrial Economics* (pp. 219–236). Macmillan Education.
6. Arruñada, B. (2012). *Institutional Foundations of Impersonal Exchange: Theory and Policy of Contractual Registries*. University of Chicago Press.
7. Arruñada, B., & Garoupa, N. (2005). The choice of titling system in land. *The Journal of Law and Economics, 48*(2), 709–727.
8. Atran, S., Chase, A. F., Fedick, S. L., Knapp, G., McKillop, H., Marcus, J., & Webb, M. C. (1993). Itza Maya tropical agro-forestry [and comments and replies]. *Current Anthropology, 34*(5), 633–700.
9. Babcock, B. (2015). Geographical indications, property rights, and value-added agriculture. *Iowa Ag Review, 9*(4), 1.
10. Baland, J. M., & Robinson, J. A. (2008). Land and power: Theory and evidence from Chile. *American Economic Review, 98*(5), 1737–1765.
11. Baland, J. M., Moene, K. O., & Robinson, J. A. (2010). Governance and

development. In D. Rodrik & M. R. Rosenzweig (Eds.), *Handbook of Development Economics* (Vol. 5, pp. 4597–4656). Elsevier.

12. Banerjee, A. V., & Duflo, E. (2007). The economic lives of the poor. *Journal of Economic Perspectives, 21*(1), 141–167.
13. Banerjee, A. V., Gertler, P. J., & Ghatak, M. (2002). Empowerment and efficiency: Tenancy reform in West Ben□al. *Journal of Political Economy, 110*(2), 239–280.
14. Banerjee, T., & Nayak, A. (2014). Effects of Trade-Related Intellectual Property Rights on the research and development expenditure of the Indian pharmaceutical industry. *Journal of Pharmaceutical Health Services Research, 5*(2), 89–94.
15. Banner, S. (2002). Transitions between property regimes. *The Journal of Legal Studies, 31*(S2), S359–S371.
16. Barlow, J. P. (1997). The economy of ideas: Everything you know about intellectual property is wrong. In A. Moore (Ed.), *Intellectual Property: Moral, Legal, and International Dilemmas* (p. 359). Rowman and Littlefield.
17. Barro, R. J. (1991). Economic growth in a cross section of countries. *The Quarterly Journal of Economics, 106*(2), 407–443.
18. Barzel, Y. (1989). *Economic Analysis of Property Rights.* Cambridge University Press.
19. Baten, J., Bianchi, N., & Moser, P. (2017). Compulsory licensing and innovation: Historical evidence from German patents after WWI. *Journal of Development Economics, 126*, 231–242.
20. Becker, L. C. (1977). *Property Rights: Philosophic Foundations.* Routledge & Kegan Paul.
21. Becker, S. (2005). Discovery from current and former employees. *American Bar Association.*
22. Bergeron, G., & Pender, J. L. (1999). Determinants of land use change: Evidence from a community study in Honduras. *Unpublished Manuscript.*
23. Berkes, F. (1989). Native subsistence fisheries: A synthesis of harvest studies in Canada. *Arctic, 35*, 35–42.
24. Besley, T. (1995). Property rights and investment incentives: Theory and evidence from Ghana. *Journal of Political Economy, 103*(5), 903–937.
25. Besley, T. J., & Ghatak, M. (2009). The de Soto effect. *LSE STICERD Research Paper No. EOPP008.*
26. Binswanger, H. P., & Rosenzweig, M. R. (1986). Behavioral and material determinants of production relations in agriculture. *The Journal of*

Development Studies, 22(3), 503–539.

27. Binswanger, H. P., Deininger, K., & Feder, G. (1995). Power, distortions, revolt and reform in agricultural land relations. In J. Behrman & T. N. Srinivasan (Eds.), *Handbook of Development Economics* (Vol. 3, pp. 2659–2772). Elsevier.
28. Blackstone, W. (2016). *Commentaries on the Laws of England.* Oxford University Press.
29. Blakeney, M. (2002). Intellectual property aspects of traditional agricultural knowledge. In R. E. Evenson, V. Santaniello, & D. Zilberman (Eds.), *Economic and Social Issues in Agricultural Biotechnology* (pp. 43–59). CABI Publishing.
30. Blakeney, M. (2017). Geographical indications and environmental protection. *Frontiers of Law in China, 12*(2), 162–173.
31. Boldrin, M., & Levine, D. K. (2008). Perfectly competitive innovation. *Journal of Monetary Economics, 55*(3), 435–453.
32. Bromley, D. W. (1991). *Environment and Economy: Property Rights and Public Policy.* Basil Blackwell.
33. Bromley, D. W. (1998). Property regimes in economic development: Lessons and policy implications. In E. Lutz (Ed.), *Agriculture and the Environment: Perspectives on Sustainable Rural Development* (pp. 83–91). World Bank.
34. Bromley, D. W., & Cernea, M. M. (1989). *The Management of Common Property Natural Resources: Some Conceptual and Operational Fallacies* (Vol. 57). World Bank Publications.
35. Bugbee, B. W. (1967). *Genesis of American Patent and Copyright Law.* Public Affairs Press.
36. Ciriacy-Wantrup, S. V., & Bishop, R. C. (1975). "Common property" as a concept in natural resources policy. *Natural Resources Journal, 15*(4), 713–727.
37. Clague, C., Keefer, P., Knack, S., & Olson, M. (1999). Contract-intensive money: Contract enforcement, property rights, and economic performance. *Journal of Economic Growth, 4*(2), 185–211.
38. Cleaver, K. M. (1993). *A Strategy to Develop Agriculture in Sub-Saharan Africa and a Focus for the World Bank* (Vol. 23). World Bank Publications.
39. Cleaver, K., & Schreiber, G. (1992). The population, agriculture, and environment nexus in sub-Saharan Africa. *Agriculture and Rural Development Series No. 1.* The World Bank.

40. Coase, R. H. (1960). The problem of social cost. *Journal of Law and Economics, 3*(1), 1–44.

41. Coase, R. H. (1988). *The Firm, the Market, and the Law.* University of Chicago Press.

42. Coase, R. H. (2013). The problem of social cost. *The Journal of Law and Economics, 56*(4), 837–877.

43. Coy, W. (2007). On sharing intellectual properties in global communities. In R. Capurro, J. Frühbauer, & T. Hausmanninger (Eds.), *Localizing the Internet: Ethical Issues in Intercultural Perspective* (pp. 279–288). Fink Verlag.

44. Daes, E. I. (1993). Discrimination against indigenous peoples. *E/CN.4/ Sub.2/1993/28, 28 July.*

45. Dales, J. H. (2002). *Pollution, Property & Prices: An Essay in Policy-Making and Economics.* Edward Elgar Publishing.

46. De Soto, H. (2000). *The Mystery of Capital: Why Capitalism Triumphs in the West and Fails Everywhere Else.* Basic Books.

47. Deininger, K., & Feder, G. (2009a). Land registration, economic development, and poverty reduction. In G. Ingram & Y. H. Hong (Eds.), *Property Rights and Land Policies* (pp. 257–294). Lincoln Institute of Land Policy.

48. Deininger, K., & Feder, G. (2009b). Land registration, governance, and development: Evidence and implications for policy. *The World Bank Research Observer, 24*(2), 233–266.

49. Demsetz, H. (1967). Toward a theory of property rights. *American Economic Review, 57*(2), 347–359.

50. Denevan, W. M. (1995). Prehistoric agricultural methods as models for sustainability. *Advances in Plant Pathology, 11*, 21–43.

51. Dove, M. R. (2003). The life-cycle of indigenous knowledge, and the case of natural rubber production. In A. Bicker, P. Parkes, & R. Ellen (Eds.), *Indigenous Environmental Knowledge and its Transformations* (pp. 213–251). Routledge.

52. Dove, M. R. (2005). Use of global legal mechanisms to conserve local biogenetic resources: Problems and prospects. In P. C. Bates, J. J. R. Arnold, & C. Folke (Eds.), *Conserving Nature in Culture: Case Studies from Southeast Asia* (pp. 45–63). Yale University Southeast Asian Studies.

53. Dutfield, G. (2017). Healthcare innovation and patent law's 'pharmaceutical privilege': Is there a pharmaceutical privilege? And if so, should we remove it? *Health Economics, Policy and Law, 12*(4), 453–470.

54. Eggertsson, Þ. (1990). *Economic Behavior and Institutions: Principles of Neoinstitutional Economics.* Cambridge University Press.
55. Ehui, S. K., Lynam, J., & Okike, I. (1994). Adapting social science to the changing focus of international agricultural research. In *Proceedings of a Rockefeller Foundation–ILCA Social Science Research Fellows Workshop Held at ILCA, Addis Ababa* (pp. 14–18).
56. Ekbäck, P. (2000). *Förfaranden Vid Planering Och Markåtkomst: En Rättsekonomisk Analys* (Doctoral dissertation). Institutionen för fastigheter och byggande.
57. Ekbäck, P. (2007). *Fastighetssamverkan för Utförande, Drift och Förvaltning av Gemensamma Anläggningar: Särskilt om Anläggningslagen och Lagen om Förvaltning av Samfälligheter.* Avdelningen för fastighetsvetenskap, Kungliga Tekniska Högskolan, Stockholm.
58. Ellickson, R. C., Rose, C. M., & Smith, H. E. (2022). *Perspectives on Property Law.* Aspen Publishing.
59. Farnley, S., Morey-Nase, P., & Sternfeld, D. (2004). Biotechnology—A challenge to the patent system. *Current Opinion in Biotechnology, 15*(3), 254–257.
60. Feder, G., & Nishio, A. (1998). The benefits of land registration and titling: Economic and social perspectives. *Land Use Policy, 15*(1), 25–43.
61. Feder, G., Onchan, T., Chalamwong, Y., & Hongladarom, C. (1988). *Land Policies and Farm Productivity in Thailand.* World Bank.
62. Field, E. (2007). Entitled to work: Urban property rights and labor supply in Peru. *The Quarterly Journal of Economics, 122*(4), 1561–1602.
63. Freeman, C. (1985). The economics of innovation. *IEE Proceedings A: Physical Science, Measurement and Instrumentation, Management and Education, Reviews, 132*(4), 213–221.
64. Galiani, S., & Schargrodsky, E. (2004). Effects of land titling on child health. *Economics & Human Biology, 2*(3), 353–372.
65. Galiani, S., & Schargrodsky, E. (2011). Land property rights and resource allocation. *The Journal of Law and Economics, 54*(S4), S329–S345.
66. Gardner, R., Ostrom, E., & Walker, J. M. (1990). The nature of common-pool resource problems. *Rationality and Society, 2*(3), 335–358.
67. Gerz, A., & Dupont, F. (2006). Comté cheese in France: Impact of a geographical indication on rural development. *Origin-Based Products: Lessons for Pro-Poor Market Development Bulletin, 372*, 75–87.
68. Gradstein, M. (2003). *Governance and Economic Growth* (Vol. 3098). World Bank Publications.

69. Grossman, G. M., & Helpman, E. (1991). Quality ladders in the theory of growth. *The Review of Economic Studies, 58*(1), 43–61.
70. Grossman, H. I., & Kim, M. (1996). Predation and accumulation. *Journal of Economic Growth, 1*(3), 333–350.
71. Hall, R. E., & Jones, C. I. (1999). Why do some countries produce so much more output per worker than others? *The Quarterly Journal of Economics, 114*(1), 83–116.
72. Hardin, G. (1968). The tragedy of the commons: The population problem has no technical solution; it requires a fundamental extension in morality. *Science, 162*(3859), 1243–1248.
73. Harris, J. W., & Guest, A. G. (1961). Ownership. In A. G. Guest (Ed.), *Oxford Essays on Jurisprudence* (pp. 106–147). Oxford University Press.
74. Higgs, R. (1979). Legally-induced technical regress in the Washington salmon fishery. *Institute for Economic Research.*
75. Himma, K. (2005). Information and intellectual property protection: Evaluating the claim that information should be free. *APA Newsletter on Philosophy and Law, 4*, 3–9.
76. Hohfeld, W. N. (1919). *Fundamental Legal Conceptions as Applied in Judicial Reasoning.* Yale University Press.
77. Iansiti, M., & Lakhani, K. R. (2017). The truth about blockchain. *Harvard Business Review, 95*(1), 118–127.
78. Johnson, R. N., & Libecap, G. D. (2019). Contracting problems and re□ulation: The case of the fishery. In R. Quentin Grafton (Ed.), *Fisheries Economics* (Vol. 1, pp. 345–362). Routledge.
79. Johnson, S., McMillan, J., & Woodruff, C. (2002). Property rights and finance. *American Economic Review, 92*(5), 1335–1356.
80. Keefer, P., & Knack, S. (1997). Why don't poor countries catch up? A cross national test of an institutional explanation. *Economic Inquiry, 35*(3), 590–602.
81. Kireeva, I., Xiaobin□, W., & Yumin, Z. (2009). Comprehensive feasibility study for possible negotiations on a geographical indication's agreement between China and the EU. *Supported by the EU-China IPR2 Project, 27*, 99–127.
82. Kormendi, R. C., & Meguire, P. G. (1985). Macroeconomic determinants of growth: Cross-country evidence. *Journal of Monetary Economics, 16*(2), 141–163.
83. Krattiger, A. F. (1997). Insect resistance in crops: A case study of *Bacillus thuringiensis* (Bt) and its transfer to developing countries. *ISAAA Briefs,*

2.

84. Lai, L. W. C. (1997). Property ri□hts justifications for plannin□ and a theory of zoning. In D. Diamond & B. H. Massam (Eds.), *Progress in Planning, 48*, 161–246.
85. Lanjouw, J. O., & Levy, P. I. (2002). Untitled: A study of formal and informal property rights in urban Ecuador. *The Economic Journal, 112*(482), 986–1019.
86. Libecap, G. D. (1993). *Contracting for Property Rights.* Cambridge University Press.
87. Lundgren, N. G. (1987). *Kampen om Naturresurserna: Ekonomisk Utveckling och Institutionella Förändringar i Lule Alvdal Under 700 år.* SNS Förlag.
88. Mann, A. J. (2010). A mongrel of early modern copyright: Scotland in European perspective. In *Privilege and Property: Essays on the History of Copyright* (pp. 51–66).
89. Maxwell, D., & Wiebe, K. D. (1998). *Land Tenure and Food Security: A Review of Concepts, Evidence, and Methods.* Land Tenure Center, University of Wisconsin-Madison.
90. Moore, A. D. (2012). A Lockean theory of intellectual property revisited. *San Diego Law Review, 49*, 1069.
91. North, D. C. (1990). *Institutions, Institutional Change and Economic Performance.* Cambridge University Press.
92. North, D. C., & Thomas, R. P. (1973). *The Rise of the Western World: A New Economic History.* Cambridge University Press.
93. Ostrom, E. (1990). *Governing the Commons: The Evolution of Institutions for Collective Action.* Cambridge University Press.
94. Ostrom, E., Gardner, R., & Walker, J. (1994). *Rules, Games, and Common-Pool Resources.* University of Michigan Press.
95. Ostrom, V. (2008). *The Intellectual Crisis in American Public Administration.* University of Alabama Press.
96. Ostrom, V., & Ostrom, E. (1977). A theory for institutional analysis of common pool problems. In G. Hardin & J. Baden (Eds.), *Managing the Commons* (pp. 157–172). San Francisco, CA: W. H. Freeman.
97. Palmer, T. G. (2005). Are patents and copyri□hts morally justified? The philosophy of property rights and ideal objects. *Harvard Journal of Law and Public Policy, 13*(1990), 817–866. Reprinted in A. Moore (Ed.), *Information Ethics: Privacy, Property, and Power.* Seattle: University of Washington Press.

98. Pandit, B. H., & Thapa, G. B. (2003). A tragedy of non-timber forest resources in the mountain commons of Nepal. *Environmental Conservation, 30*(3), 283–292.

99. Patino, R. (2009). Intellectual property rights and research disclosure in the university environment: Preserving the commercialization option and optimizing market interest. *Journal of the American Association for Laboratory Animal Science, 48*(2), 138–143.

100. Pigou, A. C. (1932). *The Economics of Welfare* (4th ed.). Macmillan.

101. Platteau, J. P. (2000). *Institutions, Social Norms and Economic Development.* Routledge.

102. Rasul, G. (2003). Factors influencin□ land-use chan□e in areas with shifting cultivation in the Chittagong Hill Tracts of Bangladesh (Ph.D. dissertation). Asian Institute of Technology, Thailand.

103. Rasul, G., & Thapa, G. B. (2003). Shifting cultivation in the mountains of South and Southeast Asia: Re□ional patterns and factors influencin□ the change. *Land Degradation & Development, 14*(5), 495–508.

104. Rasul, G., Thapa, G. B., & Zoebisch, M. A. (2004). Determinants of land-use changes in the Chittagong Hill Tracts of Bangladesh. *Applied Geography, 24*(3), 217–240.

105. Rockman, H. B. (2004). Nature of a trade secret. In *Intellectual Property Law for Engineers and Scientists* (pp. 25). Wiley-IEEE.

106. Romer, P. M. (1986). Increasing returns and long-run growth. *Journal of Political Economy, 94*(5), 1002–1037.

107. Rose, C. M. (1994). *Property and Persuasion: Essays on the History, Theory, and Rhetoric of Ownership.* Westview Press.

108. Saha, C. N., & Bhattacharya, S. (2011). Intellectual property rights: An overview and implications in pharmaceutical industry. *Journal of Advanced Pharmaceutical Technology & Research, 2*(2), 88–93.

109. Schlager, E., & Ostrom, E. (1992). Property-rights regimes and natural resources: A conceptual analysis. *Land Economics, 68*(3), 249–262.

110. Schumpeter, J. A. (1942). *Capitalism, Socialism, and Democracy.* Harper & Brothers.

111. Smith, R. J. (1981). Resolving the tragedy of the commons by creating private property rights in wildlife. *Cato Journal, 1*, 439.

112. Soto, H. D. (1989). *The Other Path: The Economic Answer to Terrorism.* Basic Books.

113. Spooner, L. (1855). *The Law of Intellectual Property.* Bela Marsh.

114. Thapa, G. B. (1998). Lessons learned from solid waste management in Kathmandu, Nepal. *Habitat International, 22*(2), 97–114.
115. Thiesenhusen, W. C. (1995). Land reform lives! *The European Journal of Development Research, 7*(1), 193–209.
116. Thrupp, L. A. (2002). Agricultural biodiversity: A key element of ecosystem health and sustainable food security. In L. R. Olson & S. E. Gaynor (Eds.), *Managing for Healthy Ecosystems* (pp. 317–331). CRC Press.
117. Tiwari, R., Tiwari, G., Rai, A. K., & Srivastawa, B. (2011). Management of intellectual property rights in India: An updated review. *Journal of Natural Science, Biology, and Medicine, 2*(1), 2.
118. Tornell, A. (1997). Economic growth and decline with endogenous property rights. *Journal of Economic Growth, 2*, 219–250.
119. Torstennson, J. (1994). Property rights and economic growth: An empirical study. *Kyklos, 47*(2), 231–247.
120. Tulasi, G. K., & Rao, B. S. (2008). A detailed study of patent system for protection of inventions. *Indian Journal of Pharmaceutical Sciences, 70*(5), 547.
121. von Benda-Beckmann, F., & von Benda-Beckmann, K. (2012). A functional analysis of property rights, with special reference to Indonesia.
122. Von Maurer, G. L. (1856). *Geschichte der Markenverfassung in Deutschland.* Enke.
123. Wachter, D. (1992). Land titling for land conservation in developing countries? *Divisional Working Paper No. 1992-28, Environment Department Working Paper No. 2.* The World Bank.
124. Waldron, J. (1988). *The Right to Private Property.* Clarendon Press.
125. Wiersma, L. L. (2004). Indigenous lands as cultural property: A new approach to indigenous land claims. *Duke Law Journal, 54*, 1061–1088.
126. Wilson, J., Yan, L., & Wilson, C. (2007). The precursors of governance in the Maine lobster fishery. *Proceedings of the National Academy of Sciences, 104*(39), 15212–15217.
127. Wolf Jr, C. (1979). A theory of nonmarket failure: Framework for implementation analysis. *The Journal of Law and Economics, 22*(1), 107–139.
128. Zak, P. J. (2002). Institutions, property rights, and growth. *Recherches Économiques de Louvain/Louvain Economic Review, 68*(1, 2), 55–73.

INDEX

G

H

I

L

M

Q

R

S

T

U

V

W

FSC
www.fsc.org
MIX
Papier | Fördert gute Waldnutzung
FSC® C083411

Kolibri 360 GmbH
Ferdinand-Jühlke-Straße 7
99095 Erfurt, Deutschland
produktsicherheit@kolibri360.de